DEPARTMENT OF THE ENVIRONMENT

Waste Management Paper No 1

A Review of Options

A memorandum providing guidance on the options available for waste treatment and disposal.

London : HMSO

First published 1976
Second Edition 1992

This Waste Management Paper has been produced by the Department of the Environment in consultation with the Scottish, Welsh and Northern Ireland Offices and with other interested parties.

It is intended to revise and further update this paper in the future. Any suggestions or enquiries should be addressed to:

The Department of the Environment
Wastes Technical Division
2 Marsham Street
London SW1P 3EB

WASTE MANAGEMENT PAPER No 1

Contents

List of Tables

List of Illustrations

Plate

Foreword

The Waste Management Paper Series currently comprises a set of 28 technical guidance papers on various aspects of controlled waste management. Aimed at a wide readership, they provide technical information on the treatment and safe disposal of particular waste types as well as guidelines for the licensing of disposal facilities. Other subjects covered include waste statistics and landfilling.

The series is now being revised and extended to incorporate recent initiatives and policies with respect to waste management practice, reflecting both European and British legislative developments.

Waste Management Paper No 1 was produced originally to provide an overview of waste treatment and disposal options and give guidance on evaluating the various methods available. This revision is being produced during a period of change in waste management organisation and legislation both at the UK and European level. Much more emphasis is being placed on waste minimisation and there are initiatives in the field of recycling and energy recovery. Further recognition of the need for environmentally sound, technology-based waste treatment and disposal methods is being given through the Environmental Protection Act 1990 and European Community Directives. Much reliance has been placed on landfill in the UK as a suitable technique and a substantial programme of research has supported this. However the technique has to be operated to high environmental standards and be the Best Practicable Environmental Option for the waste stream which employs it. Other methods are available which will increasingly compete with landfill as the overall cost of the landfill option rises.

Chapter One is an introduction. Chapter Two considers the scale of waste management in the UK and its costing and financial makeup. Chapter Three describes additional aspects of waste minimisation and recycling to be read in conjunction with Waste Management Paper No 28 on Recycling and Chapters Four to Eight describe the various techniques available for waste treatment and disposal. Chapter Nine provides an approach to evaluating these options. And Appendix One contains a resume of the Department's controlled waste research programme. There is a bibliography containing additional references which give further background reading.

CHAPTER 1

Introduction

Background

1.1 Over fifteen years have elapsed since the publication of the first Waste Management Paper No 1 in 1976. That review of the options for disposal was set in a period of change and development in the practice of waste disposal. This followed on from the implementation of the 1974 Control of Pollution Act and the establishment of Waste Disposal Authorities which, together with local government reorganisation in 1974 (1975 in Scotland), provided a framework for waste disposal to be conducted on a more strategic basis than hitherto.

1.2 The need for systematic studies of the methods of waste disposal had also been recognised and the Department of the Environment undertook major programmes of research to investigate the behaviour of wastes in landfill sites to try to understand the underlying mechanisms governing landfill disposal - a method itself given a new basis following the Sumner Report of 1971 - and further studies on methods of waste handling and treatment.

1.3 Similarly this Waste Management Paper has been produced at a time when environmental issues are considered to be of paramount interest and the potential and actual impacts of poor waste management practices are better understood and documented. In parallel the Government has introduced far-ranging legislation to improve environmental standards in the UK and has undertaken to fulfil a number of commitments put forward in the White Paper "This Common Inheritance". The Environmental Protection Act 1990 introduces the new concept of Integrated Pollution Control which considers the releases of pollutants to all media: air, water and land from scheduled industrial processes. The system of waste management licensing has been overhauled to strengthen regulatory powers and provide for the first time a requirement to judge the technical competence and the financial standing of operators of waste facilities. The Duty of Care on producers and handlers of waste, which will be a vital cornerstone of waste management in the 1990s, emphasises the responsibility of all concerned to ensure that waste is dealt with properly. Finally, the requirement to obtain a certificate of completion prior to surrendering a waste management licence will ensure that sites are left in a safe condition.

1.4 The Government is also implementing measures to strengthen the effectiveness of the land-use planning system. The Planning and Compensation Act 1991 introduces a new requirement for County Councils in England to prepare waste local plans setting out their policies and proposals for development involving the deposit of refuse or waste materials (other than mineral waste). The Act also enhances the role of the development plan, of which the waste local plan forms a part, in planning decisions.

1.5 Although the environmental effect of any proposed development has always been a major factor in development control, since 1988 there has been a specific requirement for an assessment of these effects to accompany planning applications for many types of waste disposal activity.

1.6 Many European Community Directives and some Regulations now deal specifically with wastes to re-inforce the need for high environmental standards on a Community-wide basis. Indeed, pollution is now firmly regarded as a problem that does not recognise local, national or international boundaries. Coordinated action is needed to correct the mistakes of the past as well as to provide responsible guardianship of the environment for the future.

1.7 In the UK the geological formations and the establishment of technically-based controls have enabled much reliance to be placed on landfill as a waste disposal option, but other methods have also had their part to play. The widespread use of landfill practised by Waste Disposal Authorities and the major private companies has been conducted by and large with care and without causing harm. However, there have been unfortunate cases in the past of landfill sites which have been poorly managed and have caused nuisance and pollution, often as a result of inadequate investment and charging levels. This is changing as the standards and costs of acquiring, developing, managing and monitoring landfills rise with the introduction of stricter controls. Disposal by landfill is only one of the many options available.

The purpose of this Waste Management Paper

1.8 Waste management is an essential part of daily life. Materials used by man are won from the environment and ultimately are returned to it. To continue to do so requires careful management of the way that resources are used and the products disposed of. Essentially, though, all of us are involved in waste producing activity, whether as the producer, user or consumer of products and materials. The proper management of wastes thus involves a detailed evaluation of the options for their disposal so that both the economic and the environmental costs are considered. Different circumstances will produce correspondingly different answers.

1.9 Clearly, higher standards involve costs. These costs can be offset by careful use of the best technology to minimise waste, re-use and recycle materials and to recover energy where possible; residues may be disposed of by incineration and ultimately to landfill. This pattern follows the order given by the hierarchy in the EC Council Resolution on a Community Strategy for Waste Management. The adopted Directives on hazardous waste, transfrontier movements and municipal incineration with draft Directives in progress on landfill and hazardous waste incineration reflect this strategy. This points to an increasing dependency on waste treatment technologies to achieve more efficient recovery of recyclables and energy and to deliver environmentally less harmful wastes for final deposit. It is therefore incumbent on those involved in the management of wastes to have regard to those techniques capable of delivering high standards over time. Quality matters as well as price.

1.10 This Waste Management Paper provides a general guide for those concerned with waste issues to the range of methods of treatment and disposal of controlled wastes that can be used to ensure the best techniques are matched to the types of waste arising.

CHAPTER 2

The Scale and Financial Nature of Waste Management

Introduction

2.1 Waste management in the UK is divided between three sectors: local government, which is responsible for the collection and disposal of household waste; the waste disposal sector of private industry, which deals primarily with industrial and commercial waste but which increasingly is becoming involved in the collection and disposal of household waste, and producers of waste in large quantities, which are dealt with "in-house" by individual sectors of industry themselves.

2.2 The waste management industry is subject to a number of legal controls. Some of these are quite new and relate to waste producers as well as carriers and disposers. To put this in context with the scope and scale of the industry, the first part of this chapter briefly discusses the legal framework for waste management followed by a section on the quantities of waste arising.

The legal background

2.3 The Environmental Protection Act 1990 (EP Act) improves and extends the longstanding relationship of plans and licensing introduced under the Control of Pollution Act 1974 (CoPA). It also introduces the new concepts of Duty of Care for all wastes and Integrated Pollution Control of waste from the most polluting industries. This Waste Management Paper is presented recognising that these responsibilities are going to reflect on the choice of options for the future.

2.4 The EP Act Part 1 has introduced a system of integrated pollution control for prescribed processes and substances the carrying out or use of which can be undertaken only after an authorisation has been issued. Certain processes are prescribed for central control over the release of substances to any environmental medium (Integrated Pollution Control). Other prescribed processes are designated for local control over the release of substances to the air only and are subject to the range of controls under other parts of the EP Act which includes waste management licensing under Part II.

2.5 The objectives of such an authorisation are:

(a) ensuring that in carrying on a prescribed process, the best available techniques not entailing excessive cost (BATNEEC) will be used;

(b) compliance with any directions by the Secretary of State given for the implementation of any obligations of the United Kingdom under the Community Treaties or international law relating to environmental protection;

(c) compliance with any limits or requirements and achievement of any quality standards or quality objectives prescribed by the Secretary of State under any of the relevant enactments;

(d) compliance with any requirements applicable to the granting of authorisations specified by or under a plan made by the Secretary of State under section 3(5) of the Act.

2.6 The EP Act also introduces a Duty of Care for waste, coming into force on 1st April 1992, under which it shall be the duty of any person who imports, produces, carries, keeps, treats or disposes of any controlled waste or, as a broker, has control of such waste, to take all such measures applicable to him in that capacity as are reasonable in the circumstances to:

(a) prevent any contravention by any other person of Section 33 of the Act;

(b) prevent the escape of the waste from his control or that of any other person; and

(c) secure on the transfer of the waste -

(i) that the transfer is only to an authorised person or to a person for authorised transport purposes; and

(ii) that there is transferred such a written description of the waste as will enable other persons to avoid a contravention of that section and to comply with the duty under this subsection as respects the escape of waste.

The duty does not apply to an occupier of domestic property with respect to the household waste produced on the property. Practical guidance on how to discharge the duty imposed by the Act is given by a Code of Practice.

2.7 The first step in the implementation of the Duty of Care is the registration of waste carriers introduced in the Registration of Carriers and Seizure of Vehicles Regulations 1991 under the Control of Pollution (Amendment) Act 1989. The regulations make it an offence for an unregistered carrier to transport waste and allow Waste Regulation Authorities to refuse registration in appropriate circumstances.

Land use planning

2.8 The Planning and Compensation Act 1991 (the 1991 Act) introduces a new requirement for County Councils (in England) and National Park Authorities to prepare waste local plans. Forming part of the development plan for an area, waste local plans will formulate the Authorities' policies and proposals for development involving the deposit of refuse or waste materials other than mineral waste. In the metropolitan areas and Wales, waste policies will be incorporated in unitary development plans, and district local plans respectively. A Planning Policy Guidance note on Planning and Pollution Control is in preparation. This will contain substantive guidance on waste local plans and policies and other potentially polluting development.

2.9 Most waste disposal activities require a planning application for the proposed development to be made to the appropriate planning authority. In determining planning applications under the Town and Country Planning Act 1990, local authorities must have regard to the development plan and the 1991 Act now requires that decisions on such applications must accord with the development plan unless material considerations indicate otherwise. This makes it all the more important that waste local plans/policies are prepared and kept up to date as they can be expected to figure more prominently in decisions in future.

2.10 For certain categories of development, the Town and Country Planning (Assessment of Environmental Effects) Regulations 1988, which implement Directive 85/337/EC in this respect, require formal environmental assessment (EA) to be carried out before planning permission may be granted. The Regulations set out two lists of projects. For those in Schedule 1, such as installations for the incineration or chemical treatment of special waste, EA is required in every case. For those in Schedule 2, such as installations for the disposal or incineration of non hazardous waste and waste water treatment plants, EA is required if the particular development proposed is likely to have significant effects on the environment by virtue of such factors as its nature, size or location. Where EA is required, the applicant has to prepare and submit an environmental statement with the planning application.

2.11 In the case of Schedule 2 projects, the DoE has given guidance (Joint Circular DOE 15/88, Welsh Office 23/88) on how to assess significance and indicative criteria and thresholds to assist in deciding whether EA is required. These state that installations, including landfill sites for the transfer, treatment or disposal of household, industrial and commercial wastes with a capacity of more than 75,000 tonnes a year may well be candidates for EA.

The scale of waste management activity

2.12 The management and disposal of waste involves the utilisation of significant resources which, unless used in a professional manner, can be a financial burden out of proportion with the results achieved. In addition to pollution caused by industrial processes used for the manufacture of commodities, the disposal of waste by any means may produce pollution which, unless at a sufficiently low level, places a further burden on the environment to which it is released, be it air, water or land. The greenhouse effect, a global problem, may be contributed to by waste disposal. Therefore it is in everybody's interest to ensure that the best practicable environmental option is chosen for the disposal of wastes and that the principles of BATNEEC (Best Available Techniques Not Entailing Excessive Costs) are applied where appropriate.

2.13 BATNEEC is a principle that is applied to processes that may give rise to pollution such that the stages of the process are subject to review prior to authorisation (Scheduled processes, in the case of Her Majesty's Inspectorate of Pollution or Her Majesty's Industrial Pollution Inspectorate in Scotland), to ensure that waste minimisation techniques are employed and that pollution abatement technology applied to that process is the best available.

The quantities of waste arising

2.14 Because of the division in waste management responsibilities it has been difficult to obtain statistics on the quantity of wastes handled, although it had been the intention that such statistics would be available from waste disposal plans required under Section 2 of CoPA. Further work is now being carried out to establish figures for national waste arisings.

2.15 There is very little recent information on waste arisings and disposal in England and Wales. In Scotland waste disposal statistics are collected annually by The Scottish Office. No statistics are published by the private sector and CIPFA (Chartered Institute of Public Finance and Accountancy) ceased publication of data on waste collection and disposal by Local Authorities after 1986/87. Assembling reliable data continues to pose problems because very little waste is weighed, there are no standard conversion factors for converting volumetric measures to tonnes, and there is no widely accepted classification of wastes or methodology for assembling data. This is a matter for concern and steps are being taken by the Government to rectify the situation.

2.16 Table 2.1 shows the amount of waste arising in the UK by the main source. The data are not updated on a regular basis or assembled through surveys, but are synthesised from information from a variety of sources. It can be regarded as providing a typical figure for any 12 month period in recent years.

2.17 The total of 516 million tonnes of annual waste arising differs significantly from the 700 million tonnes declared in previous editions of the Digest of Environmental and Water Statistics. The main changes are due to a re-assessment of arisings from quarrying

Table 2.1 Annual UK Waste Arisings

Waste Source	Percentage of total arisings	Arisings - millions of tonnes per annum
Controlled wastes		
Household	4	20
Commercial	3	15
Industrial	14	69
Demolition and construction	6	32
Sewage sludge *,+	-	1
Other wastes		
Agriculture	48	250
Mining and quarrying	21	108
Dredging Spoils *	4	21
Total	100	516

* Dry Weight
\+ Sewage sludge is only a controlled waste when landfilled or incinerated but not otherwise.

operations and sewage sludge. Advice from the Quarries Inspectorate is that waste from primary aggregate (sand, gravel, roadstone) extraction comprises the overburden and any contaminating materials found in the quarry. The figure is estimated at 10 per cent of production (in 1989/90) most of which is used as infill when the quarry is abandoned rather than disposed off-site. Thus the estimate in this category has been reduced by 130 million tonnes taking into account the 1989/90 estimates for allied operations. Also sewage sludges and dredged spoils are now declared on a dry weight basis to avoid spurious variations due to changes in the methods of handling and disposal which alter the water content but not the inherent amount of polluting material produced. This accounts for a further reduction of 41 million tonnes.

Other wastes

2.18 It should be noted that neither Agricultural nor Mining and Quarrying wastes are regulated as controlled wastes under the CoPA or the EP Act.

2.19 The disposal of Mine and Quarry wastes is subject to controls under the Town and Country Planning legislation. In addition the safety aspects of mineral waste tips (both solid and liquid), including their siting, drainage, design and construction, are covered by the Mines and Quarries (Tips) Act 1969 and supporting Regulations. Further guidance on disposal of these wastes is contained in the Department of the Environment's series of Mineral Planning Guidance notes (MPGs). This Waste Management Paper therefore is relevant to arisings and disposal of these wastes only insofar as they may be used in landfill sites which also take controlled wastes. In addition, as considered in Chapter 4, voids from mineral workings may constitute sites for the disposal of controlled wastes.

2.20 Agricultural wastes cover a wide variety of potentially polluting materials from manures and silage effluents to sheep dips and pesticides. Only pesticides come under the definition of controlled waste. Discharges of agricultural wastes are controlled under Section 85 of the Water Resources Act 1991 and supporting Regulations. Further guidance on the handling, storage and disposal of these wastes is contained in the Ministry of Agriculture, Fisheries and Food Codes of Practice. The Code of Good Agricultural Practice for the Protection of Water is a practical guide to help farmers avoid water pollution. It is intended that this Code will become a Statutory Code under Section 97 of the Water Resources Act 1991. Codes on the protection of air and soil are also being prepared. In Scotland, the statutory base is the CoPA, as amended by the Water Act 1989. The corresponding code covering water, air and soil is The Prevention of Environmental Pollution from Agricultural Activities.

Controlled wastes

2.21 Section 30 of the CoPA defines waste as "Controlled Waste" if it arises from household, commercial or industrial premises, the definitions being further refined in England and Wales by the Collection and Disposal of Waste Regulations 1988. These definitions are to be continued in regulations that apply throughout Great Britain under Section 75 the EPAct.

2.22 The statistics published by CIPFA for 1986/7 indicated that the total quantity of controlled solid waste handled by English and Welsh waste disposal authorities in that year amounted to about 27 million tonnes, (cf. 1990 figures above). Of this total, about half was received from collection authorities, nearly one-third from commerce and industry, about one-sixth from household ("civic") amenity sites and the rest from miscellaneous sources.

2.23 In Scotland the Districts and Islands Councils are both collection and disposal authorities. In 1989 8.7 million tonnes of controlled waste was disposed of in Scotland, recorded as 2.2 million tonnes household waste, 2 million tonnes commercial waste and 4.5 million tonnes industrial waste. Of this, 94% was disposed of to landfill, one third to Local Authority sites the remainder to private sector sites. Further details can be found in the Hazardous Waste Inspectorate Report available from the Scottish Office library.

2.24 As indicated previously (para 2.15) the accuracy of the above data must be open to question. 25 of the 39 English non-metropolitan counties and 22 of the 37 Welsh district authorities which provided the information indicated that well below half of the total waste they handled was actually weighed and 18 carried out no weighing at all. Similarly in Scotland some weighing is undertaken in only 26 of the 56 Waste Disposal Authorities.

The properties of waste

2.25 Fundamentally, waste management involves a part of the art and science of materials handling. For proper control and the most economical use of resources, it is essential to have information on the quantities involved and accurate data on the physical, chemical and biochemical properties

Table 2.2 Typical Analysis of Household Waste

(All percentages are by weight, as collected)

Classification	1935	1963	1967	1968	1969	1970	1972	1973	1977	1978	1979	1980
Screenings (1)	57	39	31	22	17	15	20	19	14	12	12	14
Vegetable and Putrescible	14	14	16	18	20	24	20	18	25	28	24	25
Paper and Cardboard	14	23	29	37	38	37	30	33	26	27	29	29
Metals	4	8	8	9	10	9	9	9	9	8	8	9
Textiles and Man-made Fibres	2	3	2	2	2	3	3	3	3	4	4	3
Glass	3	8	8	9	10	9	10	10	11	10	10	10
Plastics	-	-	1	1	1	1	2	2	5	5	7	7
Unclassified	6	5	5	2	2	2	6	6	7	6	6	4

Notes:- (1) = screenings of < 20 mm.

Due to variations in the pattern of consumer expenditure, property types, and the means used to heat them, wide variations in the proportions of constituents normally found in household waste have been measured ie:

screenings (<20 mm.)	2%-60%,	approx.
vegetable & putrescibles	20%-30%	"
paper & cardboard	8%-40%	"

of the waste, both as it is collected and on changes which take place during all stages of its handling, processing and final deposition.

2.26 To this end the Department of the Environment, in association with a number of waste collection and disposal authorities, with the help of consultants, professional bodies and contractors has, for a number of years, compiled and published tables of the analysis and average properties of typical household waste. Table 2.2 shows how these values have varied between 1935 and 1980.

2.27 Since 1980, this information has not been compiled regularly so that reliance has been placed on a more narrow range of results from specific investigations. These have been, principally, trials sponsored by DoE and Department of Energy and those undertaken by the DTI Warren Spring Laboratory (WSL), on relatively large transfer stations, civic amenity sites and reclamation and incineration plants. The data obtained, therefore, are likely to have been biased towards the types of household and commercial wastes produced in areas served by larger authorities where a need for such facilities exist.

2.28 Impending changes in the method of waste collection in some areas, particularly the more widespread use of the "wheeled bin" system for household waste, as well as seasonal variations, which are known to affect significantly both the quantity and composition of arisings, require that any practicable method of waste handling, treatment and disposal used must be flexible enough in its design to accommodate local changes.

2.29 Experience has shown the flexibility required in equipment and other facilities to cope with such variations need to be considered on a local basis and taken into account when equipment is to be specified. The scope for including local variations in figures given in a national survey is somewhat limited. It is essential, therefore, that a detailed investigation should be carried out within a particular catchment area to provide information which is meaningful for waste disposal planning purposes and for writing the specification for any facilities deemed to be necessary. The table below gives typical data prepared by WSL for an 80%:20% by weight mixture of waste arising from local authority collections and from civic amenity sites respectively. The footnote indicates the origins of the samples.

2.30 Wide regional and seasonal variations in the composition of household waste are experienced. In addition, long-term trends over the past fifty

Table 2.3 Typical Analysis of 4:1 by weight mixture of Local Authority Collected Waste and Civic Amenity Waste

Constituent	Weight % (as received)
Paper	29.2
Putrescible	19.0
Unsorted fines	8.6
Glass	8.4
Ferrous metal	8.0
Misc. Combustible	5.8
Plastic - film	4.2
Misc Non-combustible	4.0
Garden waste	3.8
Textile	3.0
Dense plastic	2.8
Wood	2.2
Non-ferrous metal	1.0

Moisture content	=	33	% by weight.
Bulk density, uncompressed	=	170	kg/m^3
Gross calorific value	=	9,260	kJ/Kg. (3,985 BThU/1b.)
Net calorific value	=	7,630	kJ/kg. (3,283 BThU/1b.)

Sample Sources:-

North:	Newcastle Merseyside Manchester Bury Doncaster Sheffield	Midlands:	Birmingham Nottingham Stoke
South-East:	Stevenage Harlow Chigwell Luton Watford	South-West:	(None)

years or so have shown a gradual increase in the quantity of waste generated per capita and in its calorific value, reflecting changes in society and industrial processes. Such factors along with many others need to be taken into account in deriving any long-term strategic planning of collection, treatment and disposal systems for household waste, particularly where extensive mechanical handling and physical, chemical or thermal processing of wastes are involved.

2.31 Similar considerations need to be made when industrial waste disposal facilities are designed and particularly when co-disposal of household and industrial waste features in a development. In such instances, there is a need also to consider the compatibility of the various waste types concerned. It is important that waste management is seen as an integral part of process design particularly in situations where hazardous waste ("Special Waste" under CoPA) may be produced which has characteristics which usually differ markedly from typical household, commercial and industrial waste.

2.32 Individual controls currently in force under the Control of Pollution (Special Waste) Regulations of 1980 and other regulations will influence the methods of collection, handling, transport, treatment and safe disposal, whether by landfill or by incineration or other means.

2.33 Many other factors have to be weighed up thoroughly before it is possible to arrive at a balanced judgement on the strategies to be followed. Such factors include transport and the inter-related cost of collection, which needs to be included to develop true disposal costs. This is particularly relevant in England, where the collection and disposal functions in shire counties are currently in separate tiers of local government. Further changes can be expected as a consequence of the EP Act which introduces a separation of waste disposal operations (in England and Wales) from direct control by local government and of the competitive tendering procedures applied to waste collection.

The financial cost of the operation

2.34 To obtain some indication of the costs involved in the collection and disposal of controlled wastes in England and Wales CIPFA data may be used as best estimates notwithstanding the reservations mentioned in para 2.15.

2.35 The cost associated with the collection and disposal of the reported 27 million tonnes of controlled waste handled by waste disposal authorities in England and Wales in 1986/7 was reported by CIPFA as being approximately £187 million, equivalent to an average net waste disposal cost, in simple terms, of £6.90 per tonne.

2.36 The corresponding net expenditure for 1986/87 by the reporting English and Welsh collection authorities was approximately £352 million, based on an estimated 17.7 million tonnes of waste collected. From these figures it can be calculated that the average local authority collection cost, again in simple terms, was £19.90 per tonne; nearly three times the reported cost of waste disposal.

2.37 In Scotland, Local Authorities reported that the disposal of 4.6 million tonnes of controlled waste in 1989/90 cost £24.2 million (£5.39 per tonne), excluding loan charges. Local Authority waste collection for the same period cost £66.5 million for 2.2 million tonnes (£29.83 per tonne).

2.38 If it is assumed that the average net cost to contractors and others in collecting and delivering the additional 9.3 million tonnes (in England and Wales) handled exclusively by them to the waste disposal authorities was about £10 per tonne, it can be calculated that the total costs associated with the collection and disposal of the estimated 27 million tonnes of waste was £632 million.

2.39 Updated to the end of the financial year 1988/89, by taking inflation into account, this would represent a combined collection and disposal cost, on the same quantity of waste, of about £750 million per annum.

2.40 Total capital expenditure by waste disposal authorities in England and Wales in this period amounted to about £40 million. Almost half of this expenditure was on landfill development and restoration, one-third on civic amenity sites and at transfer stations fitted with compaction equipment and less than one-sixth on incineration and reclamation plant. With growing concern being expressed about inadequate standards of operation and control over landfilling, and the problems associated with the control of leachate and landfill gas, the indications are that there is a need to provide more sophisticated means of control over wastes already deposited and better design and construction of sites to receive wastes. This is especially relevant in the context of the higher environmental standards that will be required and the tighter regulatory controls that will be imposed through the implementation of national legislation and as the result of EC directives and other international agreements to protect the environment. This will lead to substantial increases in capital and operating expenditure. Much of this will inevitably have to be financed from the profits of private sector based business and ultimately, the generator of the waste.

2.41 The estimated costs of collection and disposal in England and Wales (para 2.35) as the CIPFA authors themselves pointed out, are subject to some uncertainty because of the paucity of the data available. Bearing in mind the inevitable errors caused by shortcomings in the extent to which waste was and is weighed, it is probable that the average net costs of both waste collection and disposal were inconsistent and as far as so-called "direct landfilling" is concerned, were underestimates.

2.42 Also, because of the variety of methods of accounting used by local authorities, there is probably a significant imbalance between the costs of collection and those of disposal, where these are undertaken by different authorities, resulting in overall costs which cannot be regarded as optimal.

2.43 For other controlled waste arisings limited data are available in England and Wales, other than for household waste disposal. There is reason to believe that non-hazardous commercial and industrial waste disposed of to licensed landfill sites by industry constitutes an amount not less than that collected by local authorities. At a haulage and disposal cost probably averaging not less than £8 per tonne the value of the business can be estimated to be of the order of £250 million per year or more. It can be expected therefore, that the collection and disposal of controlled solid waste in England and Wales costs in excess of £1 billion annually.

2.44 The total cost of waste disposal is much higher. In arriving at the gross total cost, special waste, in-house treatment and disposal practised on a large scale in minerals extraction and processing, coal mining, the petrochemical industry, power production and the like would need to be included. Consequently, it is reasonable to estimate that the total cost associated with solid waste disposal in the UK exceeds £3 billion annually.

2.45 With such a large sum at stake it is vital that capital and revenue programmes are properly analysed with due regard for the technical evaluation, an appropriate synthesis of solutions, strategic planning and value for money.

CHAPTER 3

Waste Minimisation and Recycling

Introduction

3.1 Nearly all areas of activity involving chemical or physical processing produce waste in some shape or form. Whilst accepting that the production of waste is inevitable, the quantity of waste arising for disposal in this way should be minimised for the benefit of the environment.

3.2 Waste minimisation can be achieved in many ways, through the design of processes and selection of suitable raw materials which produce less initial waste, to the recycling and reuse of materials instead of discarding them to the waste stream. It is therefore possible to consider what opportunities there are for minimisation of waste and/or recycling at all stages in the development, production, marketing and use of products. Many wastes have a notional value and a potential for recycling but it is only those which can be re-processed into another product economically which are currently exploited and recycled. A major government priority, announced in the White Paper "This Common Inheritance", is to encourage the reuse or recycling of materials which otherwise would be thrown away and in default, to recover energy from those wastes which either cannot be recycled or for which it is uneconomic to do so. This subject is explored more fully in Waste Management Paper No 28 - Recycling.

3.3 This Chapter discusses briefly the main issues in relation to recycling and waste minimisation as described in Waste Management Paper No 28.

Terminology

3.4 There are several terms commonly used to refer to the various activities relating to the reduction of the quantity and/or quality of waste arisings prior to treatment or disposal. It is clearly better to avoid or *minimise* waste in the first place, where possible, or *reuse* it as a product. Waste minimisation and reuse are largely controlled by industry through process and product design.

3.5 *Recycling* differs from re-use because it involves a processing step. Whereas *reclamation* is used to describe the collection of materials separated from waste, *recycling* can be defined as the collection and separation of materials from waste and subsequent processing to produce marketable products.

The benefits of waste minimisation and recycling

3.6 In a consumer-based society, industry, as the producer of goods and materials which eventually end up as waste, has a major role to play. Firstly by minimising the quantity of waste arising during the manufacture of products and secondly by considering the potential of the product for recycling via the consumer who will discard the packaging and eventually discard the product as waste. Also, usually, there is scope within industry to improve the quality of its products and at the same time reduce the quantity of waste it produces.

3.7 Given the right conditions, waste minimisation and recycling should make good economic sense. It provides the means of:

* conserving natural resources;
* saving energy in production and transport;
* reducing the risk of pollution as well as saving costs in waste monitoring, treatment and disposal;
* reducing the demand for waste disposal facilities and landfill space, especially in urban areas; and
* producing goods more cheaply.

Waste minimisation in the industrial sector

3.8 British industry already recycles large amounts of waste, either in-house as part of the process or through the established reclamation industry. Some 82% of ferrous metals, 74% of copper and 66% of lead is recovered. 27 million tonnes of reusable materials valued at more than £2 billion were recovered in the UK in 1986. The overall percentage of recycled waste materials in marketed products include:

Paper and board	57%
Iron and steel	44%
Aluminium	32%
Glass	16%
Plastics	8%

3.9 Other sectors of industry where significant amounts of materials are recovered include: solvents, waste oil, textiles and rubber.

Construction materials

3.10 In some circumstances, waste and recycled materials can be used in place of newly-won primary aggregates. A study carried out in 1990 found that of a total utilisation of aggregates of about 332 million tonnes a year at the end of the 1980s in the UK, some 32 million tonnes were derived from secondary materials. The seven most important sources of these recycled materials were colliery spoil, china clay waste, slate waste, power station ashes, blastfurnace and steel slags, demolition and construction wastes and asphalt road planings. The Government is committed to increasing the use of waste and recycled materials as alternatives to primary aggregates, and further demonstration studies are planned.

Environmental Auditing

3.11 The need for a capability to investigate the possibilities for minimising waste production by improvements in the management of operations has been recognised. Many organisations are using the technique of Environmental Auditing to investigate areas where improvements can be made.

3.12 Environmental Auditing is a continuous management process which is adopted as an integral part of the company's corporate strategy. Environmental Auditing reviews a company's impact on the environment and recommends ways to reduce that impact. It involves investigations into the company's operations at all levels to recognise and implement changes which may provide cost savings related to efficiency, recycling and energy conservation, all of which are related directly to waste minimisation. At the very least it is a technique for ensuring compliance with relevant environmental legislation. Environmental Auditing is currently being considered as a topic for European legislation.

3.13 The criteria against which a particular company's performance in relation to waste minimisation can be measured include: achievements compared with environmental regulations, the advice it gives on the disposal of its products and environmental management practice being achieved by the particular industrial sector.

3.14 Keeping a close check on a company's performance in relation to recycling and to protecting the environment can improve its standing and provide increased scope for its publicity and public relations. There is evidence to show that significant savings in disposal costs can be achieved comparatively easily.

Life Cycle Analysis

3.15 Another technique which can result in waste minimisation and promote recycling is Life Cycle Analysis. This involves a "cradle to grave analysis" of particular products and their construction with respect to the raw materials used, the energy consumption necessary for manufacture, the possibility of recycling, the means of disposal and release of pollutants to the environment. One example has been an analysis of drinks containers. Their materials of construction which could be considered include: glass, metal, plastics or paper. The metal could be steel or aluminium or alloys, a range of plastics could be used and the paper could be waxed or in combination with metal or plastics. In this case, Life Cycle Analysis attempts to determine the form of packaging which overall has greatest potential for recycling and the least impact on the environment. Equally valid would be an analysis to compare the benefits of returnable containers with one trip disposal ones.

3.16 Obviously there are difficulties associated with applying the technique. For example, by-products from the manufacture of glass, metals, plastics and paper are significantly different from each other in both quantity and their properties. One suggested way of overcoming the difficulty is to apply weighting factors to potentially environmentally harmful materials present in the by-products in relation to regulatory standards concentrations acceptable in the environment. Research is being undertaken to produce reference databases on the environmental effects of the waste streams from manufacturing processes including recycling and disposal.

3.17 The economics of industrial processes always include materials costs and are related to the charge made to the customer for the product. Four categories of waste arising from the production of materials can be identified: the manufacture of products, their packaging and transport to the consumer and their disposal or recycling at the end of their useful life.

Materials production wastes

3.18 In general terms the winning and processing of raw materials for the manufacture of products takes place on a large scale. Likewise the provision of services needed such as water, electricity and gas also involve large scale processes. While the quantities of waste produced in relation to the unit quantity of product may be small, the scale of the activity means, nevertheless, that large quantities of waste are produced. Consequently, the effects on the environment from such activities can be considerable. Usually, the potential for use of the by-products elsewhere is limited.

Manufacturing wastes

3.19 The manufacture of products involves the use of materials, the production of which may have already caused some environmental harm. The choice of materials used for manufacture should take this into consideration as a longer term view and an attempt made to minimise the potential for further environmental harm in the future. The potential for recycling process residues either internally to the process or externally for another use should also be considered. Where more than one material is used in the manufacturing process the possibilities and advantages (often financial) of segregating different wastes for recycling or disposal merit attention.

Plate 1 The range of recyclable materials in domestic waste *(Harwell)*

Packaging

3.20 In many instances the amount of packaging used in the protection of products between their production and the end user can far exceed what is necessary. Again it is possible to consider the types of packaging used and consider whether other materials in less quantity, possibly also being less environmentally harmful in respect of their disposal and cost in particular, could be used. In some instances it may be possible to supply products in containers which are reusable, thus obviating waste production and disposal by the customer.

3.21 Packaging serves a cosmetic as well as a protective function. It is estimated that between 25 and 30 percent of household waste comprises discarded packaging (some 5 million tonnes per year). While it is accepted that food hygiene requires that foodstuffs are packaged adequately, overpackaging does occur. Recent initiatives by supermarkets include encouraging the reuse of plastic carrier bags. Similar practices could be adopted more widely. However, the variety of different types of plastics used for packaging are best kept segregated if value is to be obtained from them. Mixed plastics waste is, at present, almost valueless.

Disposal or recycling of products

3.22 Many wastes can be recycled provided they are relatively free from contaminants. Where it is possible to recycle products, provided they are available in sufficient quantities, there is merit in the producer and/or supplier providing an exchange system for the old product when supplying a new one. Car batteries are a good example where most old batteries are exchanged for a new battery and the old one recycled. The lead is reclaimed and the plastic cases are recycled through the manufacture of lower grade products. Materials segregation and classification processes are fairly advanced and, for example, it is possible to fragmentise an old motor car and separate glass, plastics, ferrous and non-ferrous metals from each other. More recently, cars are being produced with disassembly in mind; for example plastic components (eg bumpers) are being labelled for easy polymer type identification.

3.23 Glass bottles for supplying beer and milk have probably enjoyed the most widespread use as returnable and reusable containers. From an environmental point of view there is merit in using returnable containers for other products. For those containers and materials which are not returnable to the supplier the provision of collection facilities aid recycling. These may include bottle banks and facilities for metal drinks cans, paper, rags, used engine oil, and the like. Similar facilities at retail/wholesale distribution centres, particularly for segregating different types of packaging materials are to be encouraged.

3.24 The UK performance in recycling household and commercial waste currently is low. Less than 5% of the 20 million tonnes per year of household waste arisings is recycled. The Government has set a target of 50% of the recyclable element of household waste to be recycled by the end of the century. In support of this a wide range of experimental projects is being sponsored in co-operation with industry, local authorities and voluntary organisations with a view to testing collection and sorting techniques and to develop markets for the separated materials.

3.25 Three main types of system resulting in the segregation of household wastes are available currently. These are:

(a) "bring systems" where facilities are provided at supermarkets and other locations visited regularly by householders, in which they may deposit recyclable wastes;

(b) "collect systems" where materials are segregated by householders into various categories for collection from the doorstep or kerbside. Section 46 of the EP Act provides collection authorities with powers to specify how this should be done and can require placement of wastes for collection into separate receptacles; and

(c) "centralised segregation" where materials are segregated into different categories after collection, usually associated with energy recovery from those fractions which are not segregated.

3.26 Recycling schemes are actively supported by industry through trade associations and a comprehensive list can be found in Waste Management Paper No 28.

Markets

3.27 Materials present in household waste which can be recycled include: paper, plastics, textiles, glass, ferrous metals, aluminium and the compostable fraction. The monetary value of the materials collected for recycling depends on a variety of factors including: the degree of contamination by other waste fractions, the demand for the product after recycling and the ability of the processor to incorporate waste materials in his feedstock. As with all markets for commodities the demand is cyclical. Whereas the glass, steel and aluminium packaging industry have so far the capacity to utilise materials for recycling and have been able to absorb an ever increasing quantity of them into its feedstock, the situation in respect of waste paper and cardboard has been extremely cyclical. The variability of the market appears to depend to a degree on whether a particular commodity has a national or international market. Laws passed in one country to require recycling of a particular commodity can have repercussions internationally.

3.28 Only a finite quantity of materials can be recycled and marketed and, in broad terms, this is related to the reprocessing capacity available and how it matches the availability of discarded products. In the UK for example, the market for coloured glass cullet is limited by the export of clear glass and significant imports of coloured bottles. If all coloured bottles available were collected, they would greatly exceed UK capacity to reuse because UK production capacity and market for glass containers is dominated by clear glass. Recycling of materials can also affect alternative beneficial use, for example if very high recycle rates for paper and plastics were achieved the viability of energy recovery from municipal waste could be adversely affected.

Research and development

3.29 A number of recycling schemes have been introduced in towns and cities in the UK utilising a variety of methods of collecting and segregating the various fractions of household waste.

3.30 Of the collection systems in operation currently, five have been evaluated and have provided the following performance indicators:

(a) all schemes have reported greater than 70% participation;

(b) all schemes have exceeded or are close to achieving the government's 25% target for recycling;

(c) one scheme (Leeds) which incorporates composting of the biodegradable fraction has achieved a reduction of 62% in the volume of household waste landfilled;

(d) the net increase in the cost of collection for recycling is approximately 1.5 times the cost of collection for direct landfill.

Each scheme covered between 3,000 and 10,000 households.

3.31 Section 52 of the EP Act introduces a concept of payments for those involved in recycling waste materials. Where a waste collection authority recycles a proportion of the waste it uplifts or any person collects waste for recycling, the waste disposal authority for that area shall pay in respect of the waste retained or collected amounts representing its net saving of expenditure. It should be emphasised that credits for recycling are only payable on material actually reprocessed (not merely separated from the waste stream) and that payments to third parties are discretionary.

Product substitution

3.32 The reduction or elimination of harmful materials or their replacement with environmentally more benign materials is being undertaken as a result of international agreements. This generally requires process or product design changes to accommodate the manufacture and use of substitute materials. For example, ozone depleting CFCs used as a blowing agent in the insulation foam in refrigeration equipment are being replaced by other gases with less potential for environmental harm. Similarly, CFCs which form the working fluid in refrigerators are being removed from redundant machines for reuse or

Plate 2 Civic Amenity site operating the 'bring' recycling system *(Harwell)*

disposal by incineration. Substitutes are now available and CFC manufacture and use is to be phased out by 1997.

3.33 Other examples include a significant reduction in the use of mercury in batteries. The production of sodium, caustic soda and hydrogen by electrolysis no longer relies on mercury as the cathode material. Organo-lead compounds used as an anti-knock agent in petrol are being replaced by other materials. PCB (poly-chlorinated biphenyl) manufacture has ceased and the material currently in use is to be withdrawn and disposed of by 1999.

3.34 Waste minimisation and recycling is expected to have a far greater role in future waste management activities. The success of schemes and policies designed to promote this depends as much on commitment to these principles as to the need to develop the infrastructure and the markets to deal with them. Improvements in the methodology, the accuracy of data collection and standards of waste treatment and disposal should also encourage waste producers and handlers to consider the options available to reduce the quantity and pollution potential of wastes for final disposal.

CHAPTER 4

Controlled Waste and Landfill

Introduction

4.1 As detailed in Chapter 2 it is estimated that about 500 million tonnes of waste are produced each year in the UK. Of this total some 379 million tonnes of waste from premises used for agriculture, mining and quarrying are not "controlled waste" in terms of the Control of Pollution Act requirements. Their treatment and disposal is regulated by other means unless pollution occurs as a consequence of the disposal method used. Industry is by far the largest generator of controlled waste for disposal and uses a variety of treatment methods which usually result in a product which can be safely landfilled.

4.2 This Chapter discusses the ways in which the majority of controlled waste arisings are currently dealt with in so far as they ultimately give rise to wastes which have to be landfilled.

The need for landfill

4.3 In whatever form it arises, an irreducible minimum amount of waste will always need to be disposed to landfill. In addition, the treatment technologies applied as part of good industrial housekeeping or as a result of compliance with an authorisation issued under Part 1 of the EP Act will give rise to quantities of wastes such as sludges, filter cakes, ash etc. for which no further treatment, other than landfill, is practicable.

4.4 Independent treatment facilities exist which can modify or treat wastes to make them acceptable for landfill e.g. by precipitation and neutralisation reactions and solids separation techniques. Controlled discharges of aqueous effluents to sewer or a watercourse may be made and ultimately a residue is produced for landfill.

4.5 In the UK, industrial wastes are sent in various chemical and physical forms for disposal to landfill sites that are licensed to accept them. There, wastes may undergo reactions within the site which serve to degrade, immobilise or detoxify components of the waste.

4.6 The subsequent methods of landfill treatment are explained in the following paragraphs.

Landfill practices

4.7 Landfilling is the controlled deposit of waste to land in such a way that no pollution or harm results. The design and construction of landfill sites therefore needs to include consideration not only of the means of deposit of the waste but also to include control in both the short and the longer term over the products of waste decomposition such as leachate and landfill gas. Landfilling in the UK has generally involved the backfilling of excavations created by minerals extraction, or raising the level of natural depressions in the landscape and low-lying ground adjoining river estuaries. Increasingly, due to a shortage of voids at locations which are suitable and acceptable in planning terms, disposal of waste in structures above surrounding ground level is being allowed. A comprehensive explanation of landfilling practice is given in Waste Management Paper No 26 and the control of landfill gas is explained in Waste Management Paper No 27.

4.8 Landfill sites fall into one of three generic types depending on the range of wastes acceptable. Historically sites have been classified as follows:

(a) Mono-disposal sites are those where only one homogenous type of waste is deposited. Such sites are used usually for in-house disposal by industry of one particular waste type such as slurries which are not easily dewatered.

(b) Multi-disposal landfills accept a range of wastes but no deliberate attempt is made to utilise the various co-reactions that will occur between the wastes. Multi-disposal landfill sites are the most numerous and are used primarily for the disposal of household, commercial and general industrial wastes.

(c) Co-disposal makes positive use of the physical, chemical and biological processes taking place within the landfill to give rise to an environmentally benign deposit. Co-disposal sites are fewer in number but are used for the disposal of significant quantities of special wastes (hazardous wastes defined by Regulations) and difficult wastes (see para 4.9(c)).

4.9 The European Community is currently embarked on preparation of a draft directive on landfill. Negotiations on the draft are still proceeding and a final version has yet to emerge. The current version includes changes to these familiar definitions of landfill as follows:

(a) Inert waste sites, licensed to receive non-special, non-difficult wastes with no potential to harm the environment and which will not biodegrade under normal environmental conditions.

(b) Household waste sites, licensed to receive mainly household, commercial and biodegradable industrial wastes, i.e. those wastes of a nature similar to household wastes, which are defined in Section 75(5) of the EP Act, 1990. Commercial and Industrial wastes are defined in Sections 75(6) & (7) of the Act. These sites contain predominantly biodegradable wastes, i.e. non-special non-difficult wastes, which will readily degrade under the action of bacteria normally present in the landfill environment. Under anaerobic conditions in a landfill this process will produce an organically rich leachate and landfill gas.

(c) Hazardous waste sites, licensed to receive difficult and special industrial wastes. *Special Wastes* are currently defined in The Control of Pollution (Special Waste) Regulations 1980 as controlled wastes which contain, or are contaminated with, materials which are listed in the Schedule to the regulations, and by virtue of the presence of these materials are dangerous to human life as defined or have a flash point of 21 degrees Celsius or less. They include poisons, corrosives and flammables as well as prescription only medicines. *Difficult Wastes* are not defined in current legislation although they are defined in Waste Management Papers 4 & 26. The term includes those wastes not defined as "special", but which are specifically harmful either in short or long term to humans as a consequence of their chemical and toxicological properties. It also includes wastes with a potential of harm to the environment but exclude biodegradable wastes.

4.10 Co-disposal sites are used for the disposal of significant quantities of special and difficult wastes. They may be regarded as a subset of household waste sites. They are, ideally, sites containing biodegradable wastes into which controlled amounts of special and/or difficult wastes are deposited. Co-disposal makes positive use of the physical, chemical and biological processes taking place within a mature landfill to degrade, detoxify or immobilise components of the wastes. Co-disposal is discussed in greater detail in para 4.23 et seq.

Benefits and disadvantages of landfill

4.11 As a general rule landfilling, when properly conducted, controlled and monitored, and accompanied by good aftercare provisions, is a technique which offers an acceptable and economical way for the disposal of controlled waste, hence its widespread practice in the UK. Indeed, landfilling and landforming (i.e. the deposition of waste in shallow excavations which are landscaped when completed to a suitable above-ground profile) are in many instances the only practicable means of ultimate disposal readily available for some controlled wastes in certain areas.

4.12 Landfilling untreated household waste must take into account the need for the long-term control and containment of leachate and landfill gas. Since their recognition as major potential problems in the late 1970s, techniques for dealing with them - which clearly add cost to the whole operation - have been steadily developed and improved. In recent years, much research and development work initiated, sponsored and funded by the DoE and other government departments has shown that environmentally acceptable controls are possible. These control techniques are continually developed in the light of practical experience.

4.13 Containment, together with leachate treatment and the collection, controlled venting, flaring or use of landfill gas as a fuel, offer, when correctly designed and operated, practicable solutions. In some instances it may be necessary to re-excavate biodegradable waste where containment of its breakdown products cannot be guaranteed. It may be possible to carry out substantial sorting and pre-treatment to reduce or remove biodegradable matter (including timber and paper) to ensure compliance with planning and licensing conditions. In major city conurbations

Plate 3 Landfill site lining and leachate drainage *(Greater London Council)*

where haulage of large quantities of waste to distant landfill sites provides a straightforward solution, it may be economical and more environmentally acceptable to reduce the quantity of waste needing to be transported and to render it more stable by either mechanical sorting (to produce fuel and possibly compost) or by incineration in plant fitted with current standards of emission control, and possibly energy recovery.

4.14 However, the degree of pre-treatment needed for any particular waste stream to ensure that it can be landfilled in a safe and satisfactory manner depends upon a variety of factors all of which need to be carefully evaluated. The requirements to be met are highly specific to the site, particularly in respect of its planned after-use. There is no obvious across-the-board answer.

4.15 Landfill has been practised for many years and a considerable body of experience and expertise has been built up. Thus standards have improved as experience in landfill operations has increased. In future it would seem likely that the required standard of engineered containment of a landfill site will be governed by its potential to pollute ground and surface waters as outlined in the National Rivers Authority's draft Policy and Practice for the Protection of Groundwater (1991). There are some areas (inner source protection zones) where landfilling will be considered unacceptable and other sites where certain wastes will only be acceptable provided that site engineering and other operational safeguards are in place. While it may be possible to engineer such sites to accept a given range of wastes, the costs involved may be prohibitive. The trend will be towards fewer, more highly engineered and controlled sites, remote from centres of population and environmentally sensitive areas, where the possibility of environmental harm is minimised.

4.16 At the end of their working life, landfills require to be restored both to permit a beneficial use and to facilitate post-closure pollution control. It must be clearly recognised that the physical, chemical and biological processes operating within a landfill site can be expected to continue for many years after waste has ceased to be deposited. Landfill sites therefore require a programme of monitoring and pollution control, possibly over a period of several decades, until there is no longer a risk presented to people or the environment. This is to be coupled with a requirement to obtain a certificate of completion under the new EP Act controls.

4.17 The increasing standards demanded for landfilling have acted to increase costs to an extent that other technologies may, on economic grounds, provide a means by which the quantity of waste to be landfilled is reduced. (See also Chapter 3 on waste minimisation.)

4.18 Because of the demand for land around centres of population it is likely that land in more rural areas will come under closer scrutiny for waste disposal. However, as well as increased overall costs (including transport) associated with the operation of sites in rural areas, consideration needs to be given to long term post closure care and safety, particularly the need to protect groundwater and the management of landfill gas. This precautionary approach means that preference for "dilute and disperse" techniques is being replaced by a "concentrate, contain and treat" philosophy which can provide more reliable control and predictable performance of what are, in effect, very large biochemical reactors.

Waste processing

4.19 Modern landfill sites are operated to minimise their impact on the local environment whilst maximising their operational life. Ideally waste should be deposited in prepared cells. Equipment, notably steel wheeled compactors and other earth moving equipment, is then used to handle and compact the wastes to reduce the spread of litter, provide land stability and minimise void space. At the end of the working

Plate 4 Landfill waste handling vehicles (DOE)

day the deposited waste is usually covered with suitable material to minimise odour emissions and prevent infestation by flies and vermin.

4.20 Most household waste is collected and taken directly to a landfill site. However, in a number of areas, particularly where the landfill site is remote from urban conurbations, some form of pre-treatment has been found to be advantageous in conserving landfill void space and reducing transport costs. The principal pre-treatment methods used in the UK are baling, and to a lesser extent, wet and dry pulverisation.

4.21 Various types of baling machines are available which can compress household and similar wastes to densities exceeding 0.75 tonnes/m^3. Bales weighing around 1 tonne are commonly produced which are either self-supporting or sometimes need to be strapped. Baling waste facilitates its transport to the landfill area, with a saving on transport costs. The bales can then be mechanically handled and neatly stacked. Since it is difficult to compact the baled waste in the disposal area, care needs to be taken to minimise void space. Also, it is difficult for water to penetrate the bales and this may delay the onset of both aerobic and anaerobic biodegradation processes leading to lengthy site aftercare requirements.

4.22 Wet pulverisation has been used to produce a reduced volume of a more homogenous waste material which degrades more readily in landfill. Waste as collected is fed into a gently rotating inclined drum with an appropriate quantity of water. To achieve greater than 50% pulverisation, a residence time of several hours is required. Dry pulverisation relies on mechanical comminution of the wastes. Although the product can be landfilled, precautions against wind blow problems are necessary and it is often processed further and used as a feedstock for fuel-making processes or compost.

Landfilling of industrial waste

4.23 Mono-disposal and co-disposal are acceptable as means of landfill disposal for many industrial wastes in the UK. Co-disposal is not the indiscriminate landfilling of industrial waste together with household or other similar waste. It is defined in Waste Management Paper No.4 as disposal "where limited amounts of certain difficult wastes (solid or liquid) are landfilled with household, commercial or similar wastes from industrial sources, in such a way that environmental benefit is intentionally derived from processes operating within the landfill".

4.24 It is clear that during the coming decade the landfilling of industrial waste will come under increasing scrutiny. Nevertheless it has been demonstrated that properly designed, engineered and operated co-disposal landfill sites can represent a use of resources (i.e. household waste) which is not only efficient but also can provide an overall benefit to the environment.

4.25 Four major changes have affected co-disposal in the past decade:

(a) the successful treatment of leachate from co-disposal sites;

(b) the Directive on the protection of groundwater from pollution caused by certain dangerous substances (80/68/EEC);

(c) the assessment of the results of fifteen years of research (Knox 1990) ; and,

(d) the continued improvement of landfill standards and operating techniques.

Plate 5 Volume reduction by wet pulverisation in two DANO drums at Salford, Manchester
(Motherwell Bridge, Envirotech Ltd)

4.26 Nothing has had a greater impact on landfilling in general in the U.K. than implementation of the Groundwater Directive. The aquifer protection policies to be adopted by the National Rivers Authority will have two significant effects:

(a) the limitation of the disposal of industrial wastes to containment landfill sites only;

(b) restrictions on the range of wastes acceptable in disperse and attenuate landfill sites.

4.27 Greater specifications in guidance and the adoption of tighter landfilling controls are expected to be driven both by the results of research and by the additional powers provided to the Waste Regulation Authorities by the introduction of Part II of the EP Act.

4.28 A draft Directive on the landfill of waste was presented by the European Commission on 23rd April 1991. It seeks to harmonise Community landfill practice and to introduce high standards of environmental protection in landfilling Community wide. The draft Directive classifies landfills according to the wastes to be accepted at that site. It sets out the environmental protection measures required generally and at each type of site. It lists wastes unsuitable for landfilling. It outlines the requirements for applying for a site licence and delineates the general contents of a licence. It provides waste acceptance procedures, operational and aftercare procedures and has articles on civil liability, financial guarantees and the funding of landfill aftercare. It is at present only a draft, and is subject to discussion in the Council of Ministers.

Site selection for co-disposal

4.29 It has long been argued that the leachate from co-disposal landfill sites is no more polluting than that from household waste itself. Whilst this remains essentially true it must be recognised that there may be important differences between the types of leachate produced. The pollution potential of leachate produced by the degradation of household waste is dominated by the presence of biodegradable organic materials and compounds containing ammoniacal nitrogen. Whereas if the operation of a site licensed for the co-disposal of hazardous industrial wastes fails, then its leachate could conceivably contain high concentrations of heavy metals or persistent organic species.

4.30 In England and Wales the National Rivers Authority controls the implementation of the Groundwater Directive by means of its duty as statutory consultee in the licensing of waste operations. Similarly in Scotland control is through the River Purification Authorities. Implementation of the EC Groundwater Directive removes from these authorities the discretion to allow some impact by waste disposal operations on water resources, even on aquifers which presently have no planned use. Therefore, in selecting and

Plate 6 Leachate effluent treatment *(DOE)*

engineering a landfill site, unless a site lies within a non-aquifer zone and the prospective operator is prepared to accept a very limited range of wastes which have limited potential to pollute, long-term containment sites will become the main option.

4.31 Guidance on the engineered containment of landfill sites has been given by the North West Region of the National Rivers Authority (Seymour and Peacock, 1989). This provides specifications for minimum containment standards and is updated periodically. Selecting or engineering a site to provide long-term containment and protection of water resources is pointless without consideration of the suitability of any lining material or natural barrier in the context of the intended waste input and possible interactions between wastes and lining material.

4.32 An essential feature in the design of a landfill site is a water balance calculation, similar to that outlined in Waste Management Paper No. 26. The depth of leachate permissible in each cell is likely to be controlled in the site licence, the theoretical hydraulic head having been used to estimate the rate of flux of leachate through the lining material. Minimum requirements demand that no List 1 substances (as defined in 80/68/EC) should reach an aquifer and, since no barrier will be completely impermeable, there may be some areas of the country in which no List 1 substances can be deposited.

4.33 In view of the inherent stability of many List 1 substances, long term secure management of a site in which they are deposited needs to be assured.

4.34 For a site which can accept non-hazardous waste, site selection and design coupled with quality engineering and adherence to a well-founded operational plan should limit the potential to pollute groundwater. At co-disposal sites it is essential that the deposition of hazardous wastes is limited to acceptable concentrations of every hazardous component which may pose a threat to water resources.

Plate 7 Landfill gas flare *(ETSU)*

4.35 Guidance on appropriate loadings for certain materials is to be found in Chapter 7 of Waste Management Paper No 26 and in other papers in the series. It should be noted that the concentration of components given are normal maxima and may well require to be adjusted when taking account of other hazardous components present in waste. In order to achieve maximum beneficial use of the waste mass and to avoid poisoning large parts of the site, the toxic components within the waste should be limited to acceptable concentrations within that portion of the waste mass. The nature of such operations means that some high concentrations of certain components are

Plate 8 Restored landfill site *(Cumbria Land Reclamation Ltd)*

inevitable. Nonetheless, restrictions based on the principles outlined above will limit the area of a site which is thus sterilised.

4.36 Overall, therefore whilst landfill will always remain as the final place of deposit for wastes and residues either directly produced or as a product of intermediate treatment, the standards of engineering and operation of such sites are expected to rise to meet the higher standards of environmental protection. Pre-treatment to render wastes potentially less environmentally harmful and to reduce bulk, thus saving void space, and perhaps recovering energy and materials should become more prevalent as tighter controls are introduced.

CHAPTER 5

Composting and Anaerobic Digestion

Introduction

5.1 Composting and anaerobic digestion are biological processes in which organic material is broken down by the action of micro-organisms. In composting, the degradation process takes place in the presence of air (i.e. aerobic conditions) and results in elevated process temperatures, the production of carbon dioxide, water and a stabilised organic residue, known as humus. Anaerobic digestion (i.e. in the absence of air) relies on different species of micro-organisms which degrade organic material in an oxygen-free atmosphere to produce a carbon dioxide/methane gas mixture ("biogas") and a stabilised residue. The biogas may be collected and used as a fuel.

5.2 A high degree of stabilisation is generally achieved in 2-3 weeks by either process route. However, if the humus produced is to be used as a soil conditioner or as a fertiliser, a further stage of "curing" by slow aerobic degradation for several weeks generally is necessary to destroy any microbial by-products produced during the composting which may be toxic to plants (phytotoxins).

5.3 Both composting and anaerobic digestion are commonly used for the treatment of agricultural wastes, particularly farm slurries, and for sewage sludge. Also, both methods offer suitable processes for dealing with the putrescible organics fraction which constitutes about 35% of household waste. It is the putrescible fraction which is primarily responsible for gas production and the pollution load in the leachate in landfill sites and can be a reason for inefficient combustion in household waste incineration plant.

5.4 Humus produced from household waste is usually contaminated with heavy metals, residual glass, plastics and other materials the presence of which may reduce its utility as a soil conditioner or fertiliser. Consequently, such low grade humus may be restricted in its application and have use only as landfill cover material. However, modern composting screening technology coupled with the widespread introduction of bottle banks can greatly reduce glass contamination, and other separation regimes can separate particular metal contaminants, such as those contained in batteries, from the main waste stream.

5.5 Civic amenity wastes and household organic wastes which are separated at source can produce a higher quality compost. Several European states already segregate waste, a practice which is also being considered in pilot schemes in the UK. At Edmonton some 2500t/a of various grades of compost are already being produced from garden wastes brought to civic amenities sites and from city parks. Similar schemes are being planned to treat forestry wastes.

Composting

5.6 Composting requires careful selection and preparation of the waste if the process is to be effective. For example, wastes from parks and gardening, industrial food processes and crop residues from agriculture lend themselves to this process. The waste material needs to be ground down to a particle size of less than 50 mm to provide adequate surface area for microbial attack, and the moisture content should be in the range of 55% to 65% by weight. The microbial activity (being exothermic) generates heat and close control

Plate 9 Simple DANO composting plant at Penedes Y Garaf, Spain *(Motherwell Bridge, Envirotech Ltd)*

over aeration, temperature and moisture content is required.

5.7 Three methods of composting have been developed: Windrowing, in which the waste is placed in elongated lines some 2 metres high and uses mechanical turning to aerate the composting waste; forced aeration systems, in which a static mound of the waste is aerated either by air blowing or vacuum induction; and in-vessel composting in which an enclosed reactor system is used to allow close control over temperature, moisture content, and rate of aeration.

5.8 The first two systems have gained some application as a means of domestic waste treatment in Europe and the USA. In the first method, it is necessary to have sufficient process control to guard against excessive odour generation and the release of aerosol and particulate emissions which can lead to potential health hazards during mechanical turning. In the forced aeration system, however, ventilation may be vacuum induced allowing any emissions produced to be contained and treated. Notwithstanding this, in-vessel composting may well become the preferred option despite the relatively high capital cost involved. A combined approach, static pile or in vessel treatment followed by windrowing is becoming more commonplace in Europe.

Anaerobic Digestion

5.9 Anaerobic digestion always requires an engineered vessel approach due to the need to operate under strict anaerobic conditions. Although the process generates little heat compared to composting, facilities for the control of temperature are required.

5.10 Although moisture content is not as critical as it is with composting, control of moisture content and matching its range with the engineering design of the plant is still essential. The principal experience to date has been with waste streams such as sewage sludge, farm slurries and a number of industrial effluents having a high (>90%) moisture content. However, pilot and commercial plants abroad, such as Valorga in France and DRANCO in Belgium, have shown that moisture content up to around 60% (i.e. with no visible free water in the organic paste) can be successfully digested. The process, therefore, is likely to offer an economic solution for the treatment of segregated household waste since it results in relatively low capital costs and minimises the production of liquid effluents from the process. On the other hand, the process also offers the opportunity of combining suitable liquid effluents with the organic fraction in the waste to produce a low solids content paste which would be pumpable using existing designs of plant and provide a common treatment/disposal route for additional wastes and materials.

Plate 10 DANO composting fines *(ETSU)*

Environmental considerations

5.11 Treatment plants based on anaerobic digestion and in-vessel composting methods process the waste under tightly controlled conditions. Consequently, there are considerable environmental benefits offered by these processes in comparison with landfill. These include much enhanced control over gas, leachate and odour emissions. Also the biogas produced by the processes represents about 25% of the energy content of the waste. This gas can be utilised as a fuel on-site and/or to generate electricity and heat for export off-site. The heat generated during composting can offer opportunities for its utilisation in local space heating schemes. However, as with all treatment processes, an end product is a residue which may have to be landfilled. Although largely stabilised from a biological viewpoint with a much reduced potential for gas and leachate generation, the residues cannot be considered inert.

Costs

5.12 Current landfilling costs range from around £5 to £15 per tonne for household waste. The processing cost associated with composting and anaerobic digestion plants might be around £25-30 per tonne. However, it is possible to save on the cost of transport by siting the plant near the source of the waste. In regions where local authorities pay high transport and landfilling costs, treatment plants suitably located near to centres of population could offer a competitive option, particularly where surplus heat (albeit low grade) could be utilised and high quality humus is produced. However, modern composting and anaerobic digestion technologies for the treatment of household wastes need to be demonstrated successfully in the UK - both technically and commercially.

Outlook

5.13 Around 20 municipal composting plants were constructed in the late 1960s but all of these have since closed for technical and economic reasons. Currently in the UK, composting and anaerobic digestion processes are operated chiefly on a small scale in specialised applications eg. for food processing waste. There are at present no commercial-scale plants treating household waste. Rather, these methods offer promising technologies for application to selected waste streams. Work is required to demonstrate the technology and to decrease costs. Four principal developments may be considered:

(i) an independent system, treating small scale waste arisings (up to 50,000 tonnes/annum), primarily to stabilise the organic matter prior to landfilling, but with the possibility of materials recovery (for recycling) from the non-organic fractions;

(ii) an integrated system treating the organic residues from refuse-derived fuel production in centralised waste treatment facilities;

(iii) a complementary option to manage the organic waste stream where a substantial amount of clean recyclable materials is taken out for recycling, and;

(iv) a solution to diverting organic waste from civic amenity waste (which generally totals around 20-30% of total household waste), by provision of a separate container at the collection site.

5.14 The potential of anaerobic digestion as a complementary process in the production of refuse derived fuel is one of a number of options where financial support is available from the Department of Energy through the Energy Technology Support Unit (ETSU) Biofuels Programme. It is anticipated that several such pilot schemes may be developed in the foreseeable future.

Plate 11 Valorga plant for anaerobic digestion *(ETSU)*

CHAPTER 6

Incineration

Introduction

6.1 Incineration is an established means of processing combustible wastes originating from household, commercial and industrial sources. The principal aim of the process with regard to household and commercial wastes is to reduce the volume and thereby provide significant savings in transport costs and landfill requirements. Also, it destroys the organic, biodegradable waste components thus removing those which are responsible for landfill gas and leachate generated when untreated waste is landfilled. Industrial incinerators are designed with a view to the complete destruction of the organic component of the feedstock and its associated hazardous properties. The materials treated in industrial plant can range from organic liquids and solid residues to contaminated soil. With energy recovery (covered in more detail in Chapter 7), wastes can also be used as a positive resource. Incineration is the principal route used for the disposal of clinical waste from hospitals. About 8% of household waste in the UK is processed by 34 large scale mass burn incinerators. In addition, innumerable small scale incinerators are operated as private waste "destructors". Following international agreements to prohibit the dumping of sewage sludge at sea, incineration is an option worthy of consideration.

6.2 The past decade has seen increasing international concern over the environmental impact of incineration, particularly as a source of heavy metal and dioxin emissions into the atmosphere. Detailed research in the 1980s, particularly in Canada and Sweden, established the conditions necessary to control and minimise such emissions. Since the late 1980s this has resulted in most industrialised countries introducing stringent legislation to effect controls over incineration processes. In 1989 the European Community introduced requirements for controls over municipal waste incinerators (EC Directives 89/369/EEC and 89/429/EEC). Further Directives requiring controls over incinerators for hazardous waste including clinical waste, sewage sludge and a broad range of industrial and chemical wastes, are expected or are in preparation. These set minimum performance requirements for incinerators to be adopted throughout member states.

6.3 Regulations controlling all incineration and waste combustion processes are currently being introduced under the provisions of the EP Act. The general provisions include:

(a) requirements for incineration plant to be authorised by the regulatory authorities;

(b) specific plant operating conditions (combustion temperature, gas residence time and minimum air requirements) to ensure destruction of smoke and trace organic species;

(c) requirement to use best available techniques commensurate with cost to control environmental emissions (the BATNEEC principle) and to comply with stringent emission limits for particulates (dust and fume), metals (Pb, Cr, Cu, Mn, Ni, As, Cd and Hg), and acid gases (principally hydrochloric acid and oxides of sulphur);

(d) detailed monitoring requirements for key pollutant species and of combustion

Plate 12 Edmonton incineration plant *(Harwell)*

performance indicators (including temperature, particulate emissions, oxygen and carbon monoxide concentrations);

(e) requirements for the periodic submission of operating data and monitoring results to the regulatory authorities for scrutiny.

6.4 The above list is not comprehensive. Detailed criteria have been established for the combustion of individual waste types and for different scales of operation. All new incinerators proposed after April 1 1991 in England and Wales, and April 1 1992 in Scotland, must comply with these requirements. Existing incinerators will have to be upgraded to these requirements under a given timetable; generally by 1994 to 1996 and may be subject to interim standards in the meantime. Guidance in the first instance should be sought from Her Majesty's Inspectorate of Pollution (HMIP) and Her Majesty's Industrial Pollution Inspectorate (HMIPI) in Scotland.

6.5 In general, compliance with the new legislation will require greater control over combustion conditions with tight emission limits. These limits can normally only be achieved with the incorporation of comprehensive flue gas cleaning equipment. This will add significantly to the cost of incineration, adding perhaps 50% to the capital cost of a 400,000 t/yr household waste incinerator.

6.6 It is expected that it may prove impractical or uneconomic to retrofit additional plant to older incinerators to meet the new performance requirements. Consequently, some facilities will close down. The introduction of new incineration plant to meet the higher environmental standards will then depend on the relative cost of its construction and operation in comparison and competition with other disposal options. New incinerators may need to be larger to benefit from economies of scale and incorporate energy recovery as a means of off-setting operating costs.

Incinerator design

6.7 Incineration, properly conducted, provides a secure long term disposal option for a wide range of organic materials in that they are broken down into inorganic compounds, eg. carbon dioxide, water, oxides of nitrogen, sulphur and phosphorus, whilst inorganic substances go through a dehydration or calcination process.

6.8 Good combustion relies on the 3 T's: time, temperature and turbulence. Time is measured in seconds with 2 seconds minimum residence time in the furnace in practice being considered necessary for effective combustion of any organic material. A sufficiently high temperature is vital and for chemical incinerators a minimum of about 1200°C is the norm and for municipal and less offensive wastes, 850°C. Single point temperature measurement, however, is no longer acceptable. A time temperature profile throughout the furnace is now recommended so that these figures represent the average with no dead spots or short circuiting in the furnace. To achieve this, high turbulence is required in the combustion zone.

6.9 Incinerator design depends to a large extent on the types of waste to be burned taking into account such factors as the range of wastes, physical form, calorific value, types of combustion products and quantity of ash produced. Of the wide variety of incinerators currently available, many

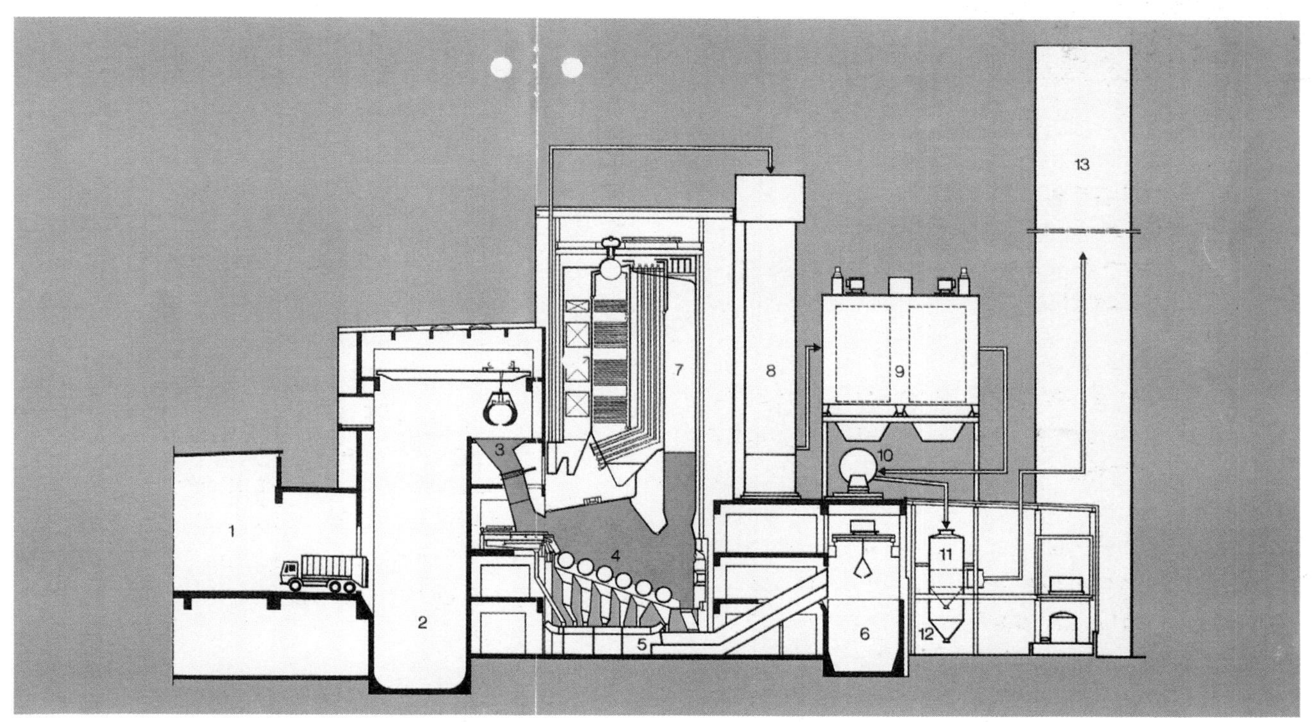

1 Tipping hall
2 Refuse bunker
3 Refuse feed
4 Roller grate/parallel-flow firing system
5 Ash extractor
6 Ash bunker
7 Steam generator
8 Spray dryer
9 Electrostatic precipitator
10 Induced draught fan
11 Scrubber stage 1
12 Scrubber stage 2
13 Stack

Plate 13 Schematic of typical domestic waste incinerator *(Motherwell Bridge Projects)*

have originated from process industries, e.g. metallurgical, cement, etc. Now each type has its niche in the overall picture.

(a) *Static chambers* are the commonest type used by waste producers who burn their own waste. Liquids are injected and solids fed directly on to the hearth or taken into the chamber on trays or drums. Modern designs incorporate an air lock system with double sets of doors on the furnace to allow solids feed without the ingress of large quantities of air to disturb the combustion conditions. These units are widely used for packaging materials and clinical wastes. With the latter the most modern units employ a series of hearths and pyrolyse the waste at around 1000°C under oxygen starved conditions. The resulting gases are incinerated. For liquids the combustion chamber may be cylindrical and can be mounted horizontally or vertically. The liquid to be burned is injected through the end plate or in a *cyclone furnace*, tangentially, using atomising nozzles and gases pass directly to a scrubber. This type has traditionally been used by chemical manufacturers to dispose of solvents, distillation residues, organic sludges and aqueous organic process liquors.

(b) *Moving hearth* incinerators were developed from coal burning plant and are mainly employed for burning municipal refuse. Again this is a chamber system the floor of which consists of a metal (steel or cast iron) conveyor on which solid materials are moved from the front to the rear by which time complete burn out is achieved. The residual ash falls off the end of the conveyor. Slots are provided in the conveyor to allow primary combustion air to pass through the material. However, municipal waste due to its heterogeneous nature requires agitation to ensure complete combustion. Thus for this type of application the conveyor is replaced by a number of moving elements which agitate and move the waste down an incline. One design consists of rollers so that the waste tumbles from one to another. Another consists of a series of bars which oscillate slowly through about 90° relative to one another, literally passing the waste along.

Plate 14 View of a roller grate from the feed chute

(Motherwell Bridge Projects)

(c) *Rotary kilns* offer a universal capability for dealing with solid and liquid wastes of all types including clinical wastes. The design essentially comprises an inclined rotating drum where the movement breaks up the waste and the speed of rotation determines the residence time. They are ideal where a long residence time is required to achieve complete burn out of the residues, i.e. for many chemical wastes. In order to comply with current combustion standards kilns are almost always followed by a secondary combustion chamber to achieve total destruction of the gaseous products.

(d) *Fluid bed* incinerators have been operated as pilot plants but as yet have not been developed to full scale. The fluidised bed comprises a bed of sand or similar inert materials which is mobilised by an upward flow of air from nozzles buried in it or through a porous plate below it. When the right flow of air is achieved the bed takes on the appearance of a boiling liquid. The bed is raised to operating temperature with gaseous or liquid fuel, then the wastes (solid or liquid) are fed in, usually at the side. The preheated sand has considerable thermal inertia and provides for uniform temperature conditions which frequently allows efficient combustion at lower temperatures than would be required in other types of incineration. In practice, the bed conditions can only cope with a uniform feed material and cannot tolerate solid feed with widely differing size fractions or densities. The technique has potential for burning organic liquids, acid tars and sludges. One application which is being brought to commercial scale in the UK is for the combustion of sewage sludge cake which has a low calorific value and is only just autothermic.

6.10 There are a number of types of process plant which can be modified to burn wastes as a secondary fuel or in combination with existing fuel supplies. The cement industry which uses kilns with long residence times compared to incinerators has experimented in the past with shredded municipal waste and tyres. Some organic solvents can be burned in industrial boilers. These are discussed in Chapter 7.

Household waste incineration

6.11 Systematic incineration of household waste was pioneered in the UK in the late 19th Century, initially as a means of sanitary disposal in urban areas but subsequently also as an opportune means of electricity production. Indeed, by 1912 some 76 plants in the UK generated power from waste in this way.

6.12 The 34 incinerators currently operated in the UK are all of the mass burn type, accepting raw waste without any pre-processing. They are therefore large and of rugged construction and incorporate substantial mechanical components to handle the raw waste and ash residues. Most of the incinerators are operated with a throughput of 6 to 10 tonnes per hour. Multiple hearth units are operated in parallel where greater plant throughput is required and to allow for planned maintenance.

6.13 The incinerators were all built by local authorities between 1968 and 1976 as a result of contracts placed before the reorganisation of local government in 1974 (1975 in Scotland) consequent upon the Local Government Act 1972 (Local Government [Scotland] Act 1973). This also predates the start of disposal licensing under the Control of Pollution Act 1974. The plants were built on very tight budgets using the lowest cost option, often by contractors who needed to license the technology from overseas. Only five of the facilities recover energy: Coventry, Edmonton (London), Jersey, Nottingham and Sheffield. Most have experienced operational difficulties and several have required major modification. The 1972 Local Government Act also reorganised waste disposal and made it a county responsibility, thus providing access to rural landfill sites which in many instances could be operated at lower cost. The consequence of these technical and economic factors is that no new household waste incinerators have been commissioned in the UK in the past 15 years.

6.14 As a result of the construction carried out in the 1970s some 7% of UK household and commercial waste (approximately 2.3 M tonnes) is incinerated, about a quarter of it in plants provided with energy recovery facilities. A significant proportion of this capacity will be lost as plants are closed where it is not economic to implement the modifications necessary to meet the new performance requirements.

European experience

6.15 In contrast, incineration has continued to be developed in continental Europe and makes a significant contribution to the means of domestic waste disposal in Denmark (60%), France (35%), Germany (35%) and the Netherlands (30%). In these countries, modern plants are generally equipped with heat recovery, often being integrated with district heating or combined heat and power schemes, and operate reliably with over 85% availability. Environmental concerns have stimulated the development of comprehensive emissions abatement technologies, particularly from German and Swedish manufacturers. Such systems are capable of routinely exceeding the current EC standards for incineration, thus enabling tighter national standards to be set.

6.16 Based on continental experience, mass burn incineration to high environmental standards must be considered a proven technology. However, in UK terms it provides a relatively high cost disposal route, typically in excess of £30/tonne compared to average UK landfill costs of less than £15/tonne. The cost of incineration may still be competitive for urban conurbations, particularly where landfill capacity locally is limited, necessitating the provision of transfer facilities and long-haul bulk transport to rural landfills. In such cases, the weight and volume reduction achievable by incineration may reduce the overall cost of waste disposal. At the same time the introduction of more stringent environmental standards for landfilling and requirements for the provision of long term aftercare can be expected to increase the cost of landfilling significantly. Thus the narrowing difference in cost between the two methods will improve the economic incentive for incineration.

Energy recovery

6.17 Energy recovery from the operation of an incinerator can provide a significant economic benefit. The 20 million tonnes of household waste disposed of annually in the UK could provide an energy value equivalent to 8 million tonnes of coal. Waste is regarded as a renewable energy resource and electricity generated from it can attract premium prices in England and Wales under the Non-Fossil Fuel Obligation (NFFO) introduced under the 1989 Electricity Act. This has stimulated renewed interest in household waste incineration. Proposals for several new plants have been made since the NFFO was introduced. The NFFO has also provided an incentive to upgrade existing incinerators to comply with European environmental standards and incorporate facilities for electricity generation at the same time. The NFFO scheme only extends to England and Wales at present.

Incinerator ash

6.18 The residual ash resulting from the incineration of household waste can be processed to recover ferrous metal but otherwise has little residual value. It has found limited use as a low grade aggregate in the construction industry but, generally, is disposed of as a cover material for landfills. The ash, particularly the deposits removed during flue cleaning operations, can contain heavy metals at high concentration and may contain organic residues. Concern over the toxicity of constituents in the ash has led to the introduction of national legislation by several countries to control its disposal. Generally, disposal in dedicated landfills is required. The

Plate 15 Industrial waste incinerator with scrubbing equipment in foreground

(Rechem Environmental Services)

major incinerator manufacturers are therefore investigating alternative post-treatment technologies for incinerator residues, including solidification and vitrification processes to form stable by-products suitable for construction use.

6.19 One approach proposed by Neutralysis Pty of Australia is to pelletise shredded household waste with clay. It is claimed that, when incinerated at high temperature, a stable aggregate is produced in which the ash and heavy metals are retained in an insoluble form. It is proposed that this material could compete with blast furnace slag as a light weight construction aggregate. Several local authorities have expressed initial interest in the technology. The suitability of this process as an energy from waste and disposal option is to be assessed by ETSU on behalf of the Department of Energy.

6.20 Fluidised bed technology also offers the potential for combining a sorbent within the combustion process to produce a stable aggregate product. A number of manufacturers are proposing such technology but, as yet, no large scale municipal waste based systems have been installed in Europe.

Hazardous waste

6.21 It is estimated that there are in the region of 70 incinerators operated by waste producers dedicated to the disposal of hazardous wastes, generally handling a specific range of organic process liquors, solvents and distillation residues.

6.22 "Merchant" incineration companies treat waste generated by other people and as a result burn a wide range of materials including liquids, slurries and solids; organic and aqueous substances; drugs, pesticides, PCBs and halogenated solvents including CFCs. Wastes are blended to give uniform characteristics (calorific value, chlorine content, heavy metals and ash) to match plant design whilst avoiding dangerous reactions between wastes. For example aqueous and other waste with a low calorific value will, where possible, be blended with high calorific value solvents to give self supporting combustion.

6.23 As waste producers develop strategies for waste minimisation, residues for disposal are becoming more difficult to handle. Solids and sludges resulting from primary treatment processes to reduce waste volume are replacing the liquids previously sent for incineration. As a result rotary kilns which can burn solids as well as liquids are becoming the standard technology for merchant incinerators.

6.24 The capital cost of a rotary kiln incinerator with a capacity for around 40,000t/a of waste is between £20-£30 million on a green field site. Operating costs vary widely with the toxicity and calorific value of the waste, usually in the range £150 - £3000/t, the higher costs being for difficult compounds like PCBs. A significant proportion of cost is directed to the provision, operation and maintenance of flue gas cleaning equipment to remove combustion by-products. Of particular concern is the potential for the formation of dioxins from incompletely combusted chlorinated organic material in the 280°C-430°C temperature

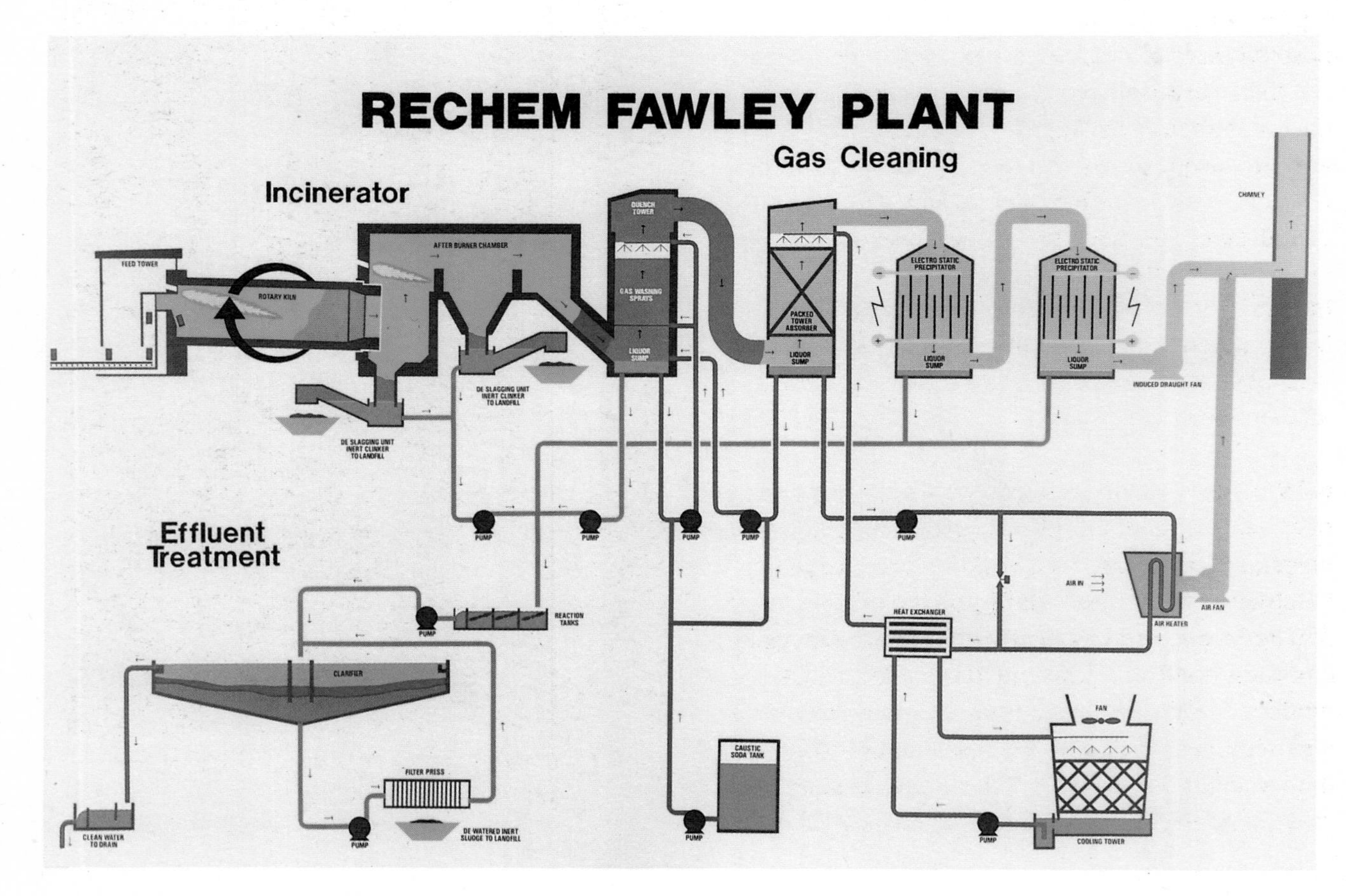

Plate 16 Industrial waste incinerator flow diagram *(Rechem Environmental Services)*

region. Thus the furnace gases are rapidly quenched to bring them below this critical range. As a result of this requirement it is unlikely that there is potential for waste heat recovery from incinerators handling chemical wastes.

Clinical waste incineration

6.25 Hospitals generate a range of wastes from that akin to household waste to infectious surgical materials. Guidance on the management and disposal of such wastes was published by the Health and Safety Commission in 1982, by the Department of Environment in Waste Management Paper No 25 and, most recently, by the London Waste Regulation Authority in 1989. All the above guidance is non statutory. Clinical waste is defined as industrial waste by the Collection and Disposal of Waste Regulations 1988. (The Regulations do not extend to Scotland.) Generally, the different waste types should be segregated within the hospital prior to despatch to an appropriate facility for their disposal. Non-clinical wastes, including uncontaminated packaging materials, paper, etc. are disposed of by private waste contractors. Clinical waste, which also arises from clinics, dental and veterinary practices, general practices, etc., has historically been disposed of by incineration at plant located on hospital sites. It is estimated that some 150,000 tonnes of such waste are disposed of annually at some 900 on-site incinerators.

6.26 Considerable concern has been expressed over the efficiency and standard of operation of hospital incinerators. The majority are old, operate batchwise with rated throughputs of less than 350 kg/hr and are equipped with only rudimentary emission control equipment incapable of meeting modern standards. In addition, Crown Immunity was removed in April 1991. It is anticipated that the overriding majority of these incinerators will close as replacement capacity becomes available.

6.27 Several types of organisation and ventures are seeking to provide replacement capacity. These include: hospital groups wishing to provide a regional facility; joint ventures between Health Authorities and incinerator manufacturers and/or private waste disposal contractors, and private waste disposal contractors who are now operating clinical waste incineration plant to service a local area. These incineration schemes are the first of their type submitted to Local Authorities for planning permission for over a decade and amongst the first new industrial processes to be presented for authorisation by the regulatory authorities under the EP Act.

6.28 A range of incineration and flue gas cleaning technologies has been proposed, generally sized to provide a throughput of about 1 t/hr.

Plate 17 Clinical waste incinerator with rotary kiln, 1/2 t/hr capacity
(Motherwell Bridge Envirotech Ltd)

Plate 18 Incinerator control room *(Rechem Environmental Services)*

Despite the attractiveness and economic potential, most of the schemes do not include provision for energy recovery in the first instance. However, discussions indicate that it is practicable to retrofit such facilities and it can be expected that such an option will be exercised once adequate plant operating experience has been gained.

Outlook

6.29 The rapid development taking place in the clinical waste incineration market can provide important lessons for waste incineration and energy from waste in general. Several instances have already highlighted uncertainties in achieving planning permission and designs of combustion appliances and/or flue gas cleaning systems capable of achieving full compliance with legislative environmental requirements. As was experienced during the phase of household waste incinerator construction in the late 1960s and early 1970s, significant reliance is being placed on technologies imported from overseas. It remains to be seen whether comparable problems concerned with earlier non-achievement of plant design specifications will occur. Important lessons about the regulatory and institutional factors affecting incineration and energy from waste may emerge. Clearly opportunities exist for the development of UK based technologies in this area in line with the development of the market for pollution abatement equipment and the use of modern treatment techniques delivering high standards for disposal and control of the wastes and residues.

CHAPTER 7

Energy from Waste

General

7.1 Estimates indicate that the household, industrial and agricultural wastes generated annually in the UK have a potential energy value equivalent to some 30M tonnes of coal (around 10% of the UK's primary energy requirements). Government policy is to view such wastes as a renewable resource and, wherever environmentally acceptable, to place their utilisation alongside materials recovery as a preferred waste disposal option.

7.2 In particular, the displacement of coal and other fossil fuels by wastes of organic origin is considered beneficial in terms of reducing the nett CO_2 balance. Harnessing landfill gas can also reduce the methane contribution to the greenhouse effect.

7.3 Using systems and techniques either being demonstrated or under development, around 50% of this potential could be economically attractive at currently available energy prices. At present only a fraction of this potential is utilised, equivalent to less than 0.5M tonnes of coal,

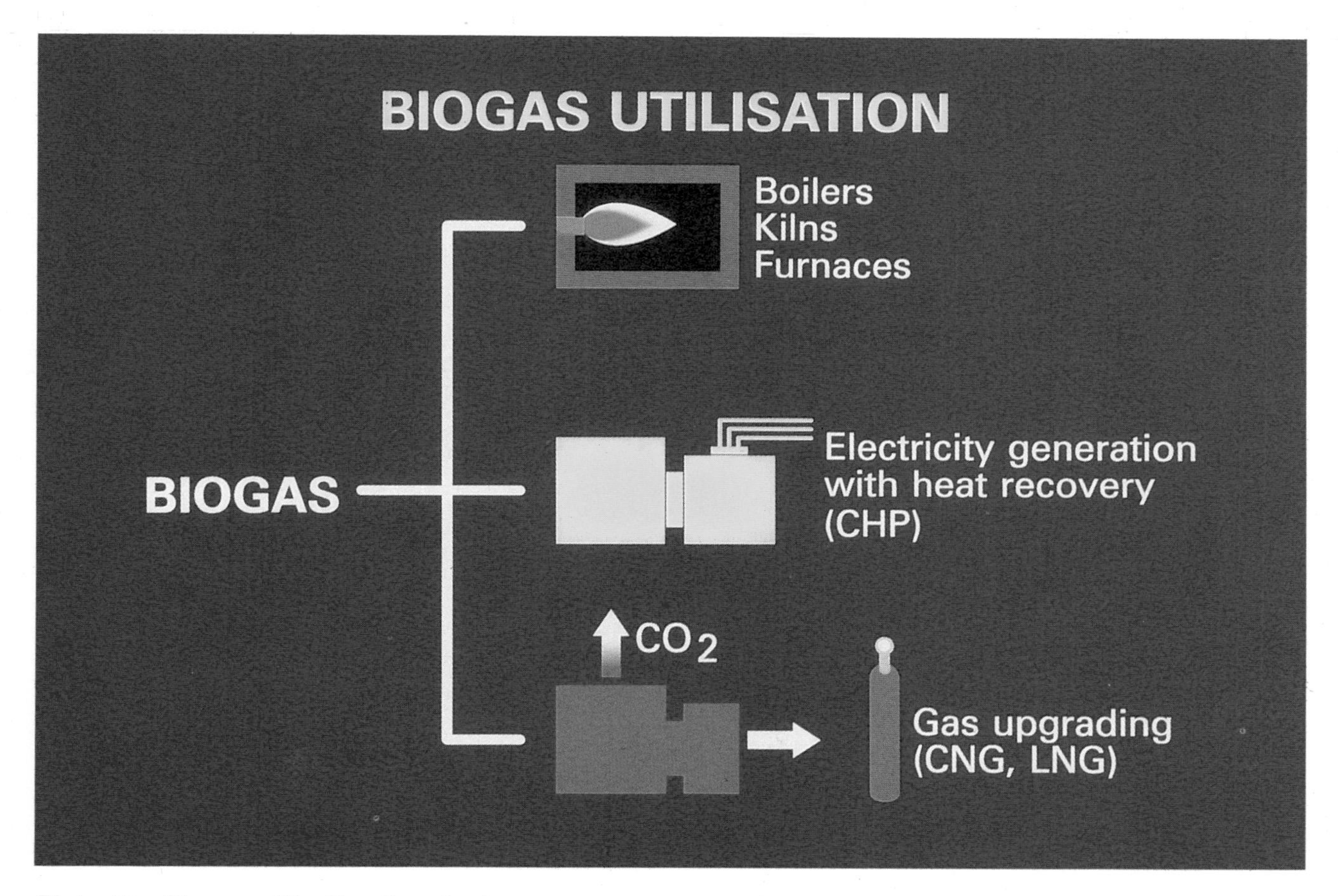

Plate 19 Biogas utilisation flowchart *(ETSU)*

predominantly from some energy recovery from municipal waste incineration and landfill gas recovery. Estimates of the currently available resource and predominant disposal routes are given in Table 7.1.

7.4 Technologies for exploiting this resource are discussed elsewhere in this paper. They include direct incineration, thermochemical treatment (pyrolysis or gasification) and anaerobic digestion, either in-vessel or through landfill gas abstraction. Energy can be recovered in the form of heat, electricity or higher value gaseous or oil products, depending on the conversion route employed.

7.5 The utilisation of wastes as fuel, however, is not always straight forward. The characteristics of waste and waste derived fuels (including bio-gas and bio-oil) differ markedly from those of conventional fuels, frequently necessitating significant modification of the utilisation system and potentially higher maintenance costs in order to achieve adequate performance. Thus the potential market should be assessed before embarking on a major alteration in waste treatment practice.

7.6 Significant parameters of waste derived fuels for combustion systems include their generally lower energy density, higher volatiles content, and frequently the presence of significant levels of alkali metals. These can result in unacceptable clinkering and fouling of the combustor and or heat recovery system.

7.7 Similarly, landfill gas is predominantly a mixture of methane and carbon dioxide. This unrefined mixture has a calorific value roughly equivalent to 50% that of natural gas and has been used successfully for electricity generation and as a source of heat for processes such as the manufacture of bricks (an example of co-incidental location of energy source and point of use). The gas may contain significant quantities of moisture and corrosive trace contaminants. Comprehensive fuel upgrading would be required to achieve

properties comparable with natural gas, although in many applications this is not necessary.

7.8 Recent changes to environmental legislation have had a major impact on the uptake of waste as an energy resource. As a result of the provisions of Part 1 of the EP Act 1990, fuels either comprising of, or derived from, wastes are themselves still considered as wastes and their use is subject to considerably more stringent environmental standards than conventional fossil fuels. Moreover, these requirements will extend to the smallest scales of operation. One exception to this, under the EP Act, is any gas product from the biological degradation of waste.

7.9 These requirements will severely disadvantage the use of wastes in relation to conventional fuels. Utilisation will therefore tend to larger scales of operation where the economies of scale have sustained an economic advantage.

7.10 In the past, the commercial potential for the extraction of energy from waste has been a low priority. Historically incinerators were grouped in metropolitan or built-up areas with the primary aim of disposing of waste. Indeed, incineration of wastes was generally used as a means of volume reduction or sterilisation prior to landfilling. In most cases consideration had not been given to energy recovery. Exploitation was generally undertaken opportunistically, e.g. to serve a local heating demand, although there were a few well publicised examples of early efforts to harness this energy resource.

Non Fossil Fuel Obligation

7.11 Considerable impetus has, however, now been given to the exploitation of energy from waste following the introduction of the Non-Fossil Fuel Obligation (NFFO) under the Electricity Act 1989 (England and Wales only). Under this Act the Secretary of State for Energy requires each Regional Electricity Company (REC) in England

Table 7.1 Arisings and Current Principal Disposal Routes for Wastes with Energy Recovery Potential

Waste	Estimated UK Arisings (Mt/y)	Energy Content (Mtce/y)	Current Principal Disposal Route
Municipal	20	6.6	90% to landfill, 8% incinerated, 2% other.
Industrial and Commercial	26	15.3	Mainly to landfill by private waste contractors.
Chemical			
- Solid/Liquid	0.35	0.35	0.16Mt incinerated either by specialist companies or in-house; few systems with energy recovery. Remainder landfilled with/without chemical treatment. Disposal costs £2-4k/t.
- Solvent Vapour	0.8	0.8	80% of solvent used lost as low CV vapour; limited number of sites fitted with incinerators. Some liquids recycled for reuse, either in-house or through special companies. Limited incineration of residues; few with energy recovery. Disposal costs £0.1-2k/t.
Fragmentiser Residues	0.5	0.37	To landfill (£10-40/t) on company or private waste contractor landfill sites.
Hospital	0.37	0.2	Bulk batch incinerated at hospitals, largely without heat recovery. Disposal costs up to £350/t (av. £150/t).
Meat Processing Residues/Animal carcases	1.75 (wet)	0.2	Previously processed as protein reinforcement in animal foods; concerns over BSE have resulted in disposal to landfill.
Poultry			
- Litter	1.4	0.7	Bulk to landspreading as P/K fertiliser. Legislation anticipated to control this activity. Some burnt for energy production under NFFO. Current disposal costs limited to transport: £0-0.5/k.
- Droppings	0.4 (air dried)	0.21	Bulk to landspreading. Legislation 85% dry matter anticipated to control this activity.
Mushroom Compost	0.3	0.03	Limited horticultural use. Majority dumped to landfill. Current interest in its use for land reclamation.
Sludges			
- Sewage	~1.5Mt (dry) 37.5Mm³ (wet)	0.6	40-45% soil conditioner, 30-35% sea dumped, 20-25% landfilled, ~5% incinerated. Controls on sea disposal and agricultural use will lead to other disposal methods.
- Pig Slurry	~1Mt (dry)	0.6	Storage and spreading. Legislation pending to control odour/water pollution problem. Limited work with anaerobic digestion; limited work abroad on dewatering/combustion.
- Pharmaceutical	0.02 (dry)	0.01	Currently discharged to sea or sewerage system; legislation anticipated to control this activity. Interest shown in wet air oxidation.
Tyres	0.35	0.42	Figures refer to tyres which are surplus to requirements. Landfill and surface stockpile; some incineration with energy recovery. Disposal costs up to £60/t (av. £40/t).
Wood			
- Sawmills (damp)	0.02	0.01	Uses found for 99% of this waste, hence low figure.
- Processing	0.50	0.3	Bulk to landfill via private contractors; disposal costs £15-25/t. Some on-site use as fuel for space heating.
- Demolition	1.0	0.6	Bulk to landfill via private contractors. Recycling of demolition rubble opens up opportunities for using timber residue as fuel; disposal costs limited to value of void space (£5-10/t).

and Wales to provide a set amount of electricity from non-fossil fuel sources, to be purchased at a premium price financed by a levy on electricity generated from fossil fuels.

7.12 The NFFO applies to all renewable energy technologies, and accepted schemes will receive a premium price for electricity generated until 31 December 1998. The first Renewables Order, published in September 1990, required the RECs to contract for 102 MW Declared Net Capacity (DNC). At that time the Government also announced its intention to make a further series of Orders to secure up to 600 MW renewables capacity by 1998. Accordingly, in November 1991 a second Order was laid before parliament which set an additional obligation of 457 MW DNC, with 386 MW being contributed by waste combustion or landfill gas schemes. A breakdown of the two renewables orders is given in Table 7.2.

Table 7.2 Energy from Waste Projects Accepted Under the 1990 and 1991 Renewables Tranches

	Project Type	No. of Projects	Total DNC (MW)
NFFO 1	MSW Combustion	4	41
	Poultry Litter	2	25
	Scrap Tyres	1	20
	Landfill Gas	25	36
NFFO 2	MSW Combustion	10	261
	Poultry Litter	1	9
	Scrap Tyres	1	8
	Other Waste Combustion	2	13
	Landfill Gas	28	48
Totals	MSW Combustion	14	302
	Poultry Litter	3	34
	Scrap Tyres	2	28
	Other Waste Combustion	2	13
	Landfill Gas	53	84
Cum Total	Energy from Waste Schemes (excluding sewage and slurry digestion)	74	461

Outlook

7.13 Industrial, agricultural and municipal wastes represent a potentially significant renewable energy resource. Development and demonstration of this is actively supported through Government programmes and initiatives, particularly through the Department of Energy's Biofuels Research Programme. The options currently being considered are:

(a) Mass burn incineration of municipal and general industrial wastes is well established but, under stringent environmental standards, may only be economically viable where sufficient quantities of waste are available to support large scale schemes.

(b) Resource recovery, involving the integration of centralised processing for materials recovery with energy production from the process residue, may provide an economically sound and environmentally acceptable means of waste disposal at intermediate scales of operation, typically of the order of 200,000 t/yr throughput.

(c) Small scale combustion technologies may be most constrained economically by the requirements for comprehensive emissions abatement. Opportunistic uses or applications may continue where alternative disposal routes are restricted and/or incur a high cost.

(d) Thermochemical conversion technologies are at an early stage of commercial development. However, they offer significant potential to exploit biomass and waste fuels with a high overall conversion efficiency and an overall lower impact on the environment.

(e) Anaerobic digestion is a relatively high cost technology, but has the capability of dealing with a wide variety of liquid and sludge wastes. It also offers significant potential for integration with resource recovery for the treatment of the putrescible fraction of municipal wastes.

(f) Landfilling. Tightening standards for landfilling, including the Duty of Care introduced under the Environmental Protection Act, will increase landfill costs and raise engineering standards of landfills. In turn, it is anticipated that these measures will lead to improved landfill gas abstraction and control. Under these circumstances the use of landfill gas as an energy source can offer considerable commercial benefit.

7.14 Uptake of these technologies in the long term, however, will depend on their ability to guarantee long term safe, efficient disposal of wastes in an environmentally acceptable manner.

7.15 The extent to which energy recovery from municipal waste will feature may depend on the proportion of combustibles removed from the waste stream through recycling initiatives encouraged by government policy.

CHAPTER 8

Other Waste Treatment Systems

Gasification and Pyrolysis

Introduction

8.1 Gasification and pyrolysis are related process techniques for the thermal breakdown of organic material. Historically both have been used to generate towns gas, coke and liquid feedstocks for the chemical industry from coal. As conversion processes they offer significant benefits over destruction by direct combustion when applied to highly volatile feedstocks such as wastes and biomass. The principal benefits are:

(a) potential for environmentally clean process routes with low emissions;

(b) conversion to a gaseous or liquid fuel facilitates direct conversion of the product to electricity using internal combustion or gas turbine engines. In turn, this results in higher overall conversion efficiencies when compared with the conventional combustion/ steam cycle route;

(c) conversion to a liquid fuel results in an increase of the energy density of the product compared with the waste and biomass feedstocks, thus reducing the cost of subsequent handling and transport;

(d) potential for conversion of the products to higher value materials, e.g. transport fuels and chemical feedstocks.

8.2 Table 8.1 gives a summary of gasification and pyrolysis processes that can be utilised. The technology and designs used for the process plant reactors may be similar in both cases. Generally, either fixed bed or moving bed reactors are used. The fixed bed reactor is operated in the dense phase with a high solids ratio, whereas the fluid bed reactor operates in a lean phase. Fixed bed reactors tend to use more simple technology with few moving parts. Fluid bed reactors are more complex mechanically thus requiring a greater amount of technical support but offer better control over the operating conditions.

Gasification

8.3 The process converts the bulk of the carbon present in the feedstock to gas and leaves a virtually inert ash residue in the reactor. This is achieved by partial combustion of the feedstock in the reactor with oxygen derived either from air, oxygen enriched air, pure oxygen, or steam. Relatively high temperatures are needed: 800-1100°C with air, and 1000-1400°C with oxygen.

8.4 Air gasification is the most widely used technology since a single gaseous product is formed at high efficiency and without requiring expensive and potentially hazardous air separation plants. A low calorific value gas (4-6 MJ/m^3) containing up to 60% nitrogen is produced.

8.5 Oxygen gasification produces a better quality gas (10-18 MJ/m^3 (250-500 Btu/scf)) but the use of oxygen does increase the problems of cost and safety. A medium heating value gas can also be produced by other processes such as steam gasification, high temperature pyrolysis and other more exotic technologies which avoid the need to use oxygen.

Table 8.1 Summary of Methods of Thermal Treatment

Method	Attributes
Air Gasification	Air oxidises part of feed to generate heat to gasify the rest; Product is low heating value fuel gas with up to 60% nitrogen, 4-6 MJ/m^3 ; Temperature 800-1000°C; Usually 1 bar pressure.
Oxygen Gasification	Air separation required; Gives a nitrogen free medium heating value product; Higher temperature and oxygen require better control, and enhanced safety aspects; Temperature 1200-1400°C; Up to 20 bar pressure; Better quality gas, 10-15 MJ/m^3, lower tars.
Steam Gasification	Energy supplied by steam reforming reaction which is exothermic only at high pressure, typically above 7 bar; Steam also added as thermal moderator in oxygen gasification; Temperature 700-900°C; Up to 20 bar pressure; Better quality gas, 15-20 MJ/m^3 (400-600 Btu/scf).
Pyrolysis	Indirect heating through external heating or addition of hot/inert heat transfer medium; No reagents added; Temperature 700-1000°C; Better quality gas, 17-23 MJ/m^3 (450-650 Btu/scf); High tars from pyrolysis; Hydrogasifaction (pyrolysis in a high hydrogen or carbon monoxide environment) discourages formation of oxygen rich compounds; Reagents may be added such as carbon dioxide to improve carbon conversion efficiency; or methane which suppresses methane formation and/or gives high yields of hydrocarbons; Catalysts added to non-equilibrium systems to encourage advantageous reactions, e.g. nickel for methane formation and calcium/magnesium for tar cracking; Energy efficiency 70-75% to clean cold fuel gas, up to 90% to clean hot raw gas.

8.6 The use of steam for the gasification process encourages reforming reactions and involves gasification and/or pyrolysis using external heating to drive the process. The now abandoned Oxygen Donor Gasifier process relied on chemical oxygen gasification in a horizontally circulating twin fluid bed reactor.

Pyrolysis

8.7 The pyrolysis process relies on the feedstock being heated indirectly in a sealed vessel in an inert atmosphere to produce gaseous, liquid and solid (char) products. The products and their relative proportions depend on the conditions under which the process is operated. Normally, relatively low temperatures between 600-900°C are used. Operation of the process at low temperatures favours the production of liquids whereas at higher temperatures (>750°C) gaseous products are formed. Liquids produced have a heat value in the range of 50-80% of mineral fuel oil. Fast or flash pyrolysis at temperatures above 750°C is used to maximise gas production. The gas produced is of medium heating value, 13-21 MJ/m^3 (350-600 Btu/scf).

8.8 In comparison with gasification, the pyrolysis process is not favoured for fuel gas production for the following reasons: the multiplicity of by-products formed which are difficult to handle and are not easily marketable; the lower efficiency in gas production and the technical problems of efficient heat transfer. The latter problem has been overcome by using a twin reactor system which incorporates a char combustor to provide heat for the pyrolysis and/or steam gasification process. Such systems yield a better quality gas, but at a penalty of lower overall conversion efficiency and higher cost.

Outlook

8.9 At present, the commercial incentive for using gasification or pyrolysis as a means of processing biomass feedstocks stems as much from environmental problems associated with the waste and difficulties in its disposal, as from the promotional efforts that have been pursued to demonstrate the power production potential of the technology. This is especially so when considering the disposal of urban and industrial derived wastes. These materials tend to have a low or negative value which makes them economically the most attractive feedstock materials for gasification or pyrolysis.

8.10 Thermal processing of comparatively higher value biomass materials such as wood or straw is potentially less attractive overall. However, for small scale plant the use of internal combustion engines and gas turbines fuelled by the products of thermal processing can provide a much higher conversion efficiency in power production than is achieved when using the conventional combustion/steam turbine to generate electricity. Such considerations are more relevant to materials such as wood and straw since they are derived from widely dispersed sources.

8.11 Significantly more research and development and demonstration projects to evaluate these technologies can be expected in the future. They will be carried out in parallel with developments in mass burn technology. The evaluation of alternative routes and options need not be viewed as being mutually exclusive since specific factors such as those related to availability of feedstock, its location, scale of the operation, etc. will dictate particular selections.

8.12 The products of thermal processing can be considered as potential feedstock materials for non-energy purposes eg. as synthesis gas. However, since the production plant would generally need to be large and hence capital intensive it could be considered to be at a disadvantage compared to its use for providing fuel for energy or power production. Thus, currently, electricity generation should be viewed as the lead product market for this technology.

Refuse-Derived Fuel

Introduction

8.13 Household waste can be separated mechanically into its component fractions by a process of screening, shredding and separation to yield a combustible product and fractions providing materials which may be recycled. Considerable research effort was devoted to the development of such processing in the 1970s and 1980s, both as an option to assist recycling and for reducing landfill space requirements. Originally the work was directed towards the recovery of ferrous and non-ferrous metals, the compostible fraction, glass, paper, plastics and a fuel rich fraction. However, difficulties in establishing markets for the separated materials meant that only the ferrous metal and fuel fractions were marketable and thus able to be developed successfully. The fuel fraction is called refuse derived fuel (RDF).

8.14 Two main types of RDF have been produced, coarse and densified. These are differentiated by the degree of processing required to produce them. Flock or coarse RDF (cRDF) is the fraction which remains after an initial separation of metals, glass and other non-combustible material and generally represents 50-80% of the unprocessed waste volume. Coarse RDF is more suited for direct on-site use since it cannot be stored for any considerable period of time. Densified RDF (dRDF) is produced by further processing to remove putrescible and heavy material. It comprises essentially only the paper,

Plate 20 Pebsham waste derived fuel plant *(East Sussex County Council)*

plastic and textile fraction of the original household waste (30% of the weight). When this fraction is dried and pelletised it produces a fuel which can be stored and has roughly twice the calorific value of the raw waste and an energy density roughly half that of coal.

8.15 The densified product was originally developed to compete directly as a distributed fuel, principally as a substitute for coal for use by industry. However the high fuel prices of the 1970s have not been maintained, putting dRDF at a price disadvantage. Also boiler plant generally requires modification particularly to air and fuel feed arrangements to optimise combustion, which reduces its attraction as a fuel substitute in industrial plant.

8.16 Table 8.2 lists the RDF facilities that have been built in the UK and their current status.

Coarse RDF

8.17 Coarse RDF may be burnt efficiently either alone in plant designed or modified specifically for the purpose or, typically, as a 20% admixture in a conventional coal fired plant. Both methods have been used, particularly in the USA. Firing the product in a dedicated facility can be viewed as being in competition with mass burn incineration of raw waste. However, burning cRDF offers a number of distinct advantages:

(a) the calorific value of the fuel is higher and the combustion characteristics of the fuel are more uniform;

(b) cRDF has a lower moisture and heavy metals content, thus reducing the burden on the flue gas cleaning plant necessary to limit emissions to atmosphere;

(c) ash handling requirements are greatly reduced since the feedstock contains a much lower proportion of non-combustible materials;

(d) the combustion conditions can be tailored to a particular fuel specification, thus increasing efficiency;

Table 8.2 RDF Production Facilities in the UK

Location	Completion	Throughput Tonnes/hour	RDF Product	Status
Westbury, Wiltshire	1979	8	Coarse	Closed
Grimsby, Humberside	1979	1	Coarse	Closed
Byker, Newcastle	1979	30	Pelleted	Working
Doncaster	1979	10	Pelleted	Closed
Eastbourne		1	Pelleted	Closed
Govan, Glasgow	1985	1	Pelleted	Closed
Birmingham	1985	30	Pelleted	For Sale
Merseyside	1985	20	Pelleted	For Sale
Isle of Wight	1988	8	Pelleted	Working
Pebsham, Hastings	1989	30	Pelleted	Working

(e) processing the waste provides the ability to use a more simple and less robust grate system in the furnace resulting in a significant saving in its cost and maintenance.

8.18 In the USA it has been shown that only household waste incineration facilities designed to operate at over 1000 tonnes/day cRDF provide savings in capital costs over incineration of crude wastes. With smaller throughput plant the additional cost of processing is considered not to be financially advantageous since it is not possible to make sufficient savings in the cost of the combustion and gas clean-up plant. Developments using fluidised bed technology to burn cRDF, particularly in Sweden, are said to offer significant potential advantages when used at a considerably smaller scale of operation.

8.19 Conventional incinerators may be used to burn cRDF. However, the benefits of the more simple grate designs cannot be realised. Nevertheless, the reduction in quantity of ash and improvements in calorific value of the waste can result in improved overall performance. Trials at the Sheffield incinerator and subsequently at Edmonton confirmed that pre-screening municipal waste enhances the calorific value and improves combustion characteristics in a conventional heat recovery incinerator. A front end screen has now been installed at the Edmonton incinerator to control calorific value and recover paper and board for recycling.

8.20 Most cRDF plants co-fire with conventional solid fuels. In the USA there is extensive experience using cRDF in spreader stoker fired water tube boilers and in Sweden with co-firing cRDF with peat and other biomass fuels. UK experience to date has been mixed. Extensive work was undertaken in a collaborative agreement between industry and a local authority to build and operate a dedicated 5,000 t/yr processing line to produce cRDF for co-firing in a water tube boiler. Numerous technical difficulties were encountered and resolved, both in cRDF production and its utilisation. However, institutional factors, particularly in control of the waste supply, proved insurmountable, and the plant was closed.

Use in cement kilns

8.21 cRDF has also been used as a partial substitute for coal in cement manufacture both in the UK and overseas. Cement kilns have been found to be particularly well suited for cRDF co-firing since the high temperatures, long residence times and alkaline environment ensures complete combustion and minimum pollutant carry over. A process was developed for co-firing wet process kilns with household waste collected by a local authority. The waste was delivered by the collection authority to the cement works where pre-treatment and flock production took place. The process operated satisfactorily for several years, although necessitating some turndown in cement production capacity. It is understood that the operation is currently mothballed pending negotiations on the waste supply contract.

8.22 Both these examples show that RDF can be used successfully and effectively as a substitute for other fuel. However, they highlight the need for long term economically viable back-to-back contracts for waste supply and/or fuel delivery in order to sustain and secure large scale commercial ventures.

Densified Refuse Derived Fuel

8.23 Developments in the UK have concentrated almost exclusively on the production and use of dRDF as a distributed substitute fuel. The principal reason for this is the existence of a large number of small to medium size range of coal fired boilers which serve the industrial/commercial sectors in the UK. These were considered to be the main outlets for the fuel.

8.24 With the notable exception of one dRDF plant, at least over the period 1987 to the present day, all dRDF production facilities have experienced considerable difficulties in securing outlets for the fuel produced. The early notion of dRDF producers that pellets could easily substitute for coal turned out to be optimistic given the very different properties of the two fuels. Thus when dRDF was fired on combustion systems optimised for coal burning, problems such as ash clinkering, boiler tube fouling and increased volumetric throughput of fuel had to be addressed.

8.25 The Department of Energy undertook a comprehensive R&D programme (completed in June 1991) which set out to investigate the problems associated with the combustion of dRDF on small industrial boilers. Under this programme numerous combustion trials were conducted on a variety of boiler types including those which had undergone modification and design development. In consequence, the fundamental characteristics of dRDF combustion are now well understood and the majority of the problems can be overcome by appropriate changes in boiler design and in their mode of operation, particularly with regard to sootblowing and regular and thorough maintenance.

8.26 This development has taken place against a decline in the industrial use of solid fuels in favour of the convenience of gas and oil firing.

Outlook

8.27 Waste processing was originally developed as an alternative to direct disposal. However,

Plate 21 Stockpile of waste derived fuel *(East Sussex County Council)*

markets for the recovered materials have failed to be developed sufficiently and the viability of the facilities have relied on the sale of the fuel by-product (RDF) and ferrous scrap metal only.

8.28 Sustained success has been achieved only where the fuel has been used on-site in dedicated appliances. Currently, one district heating network is fuelled in this way and two further electricity generating schemes are under construction with support of the Non-fossil Fuel Obligation (NFFO). Now it is considered that this is the primary option for production and use of RDF. Such use negates the need for densification since the RDF can be utilised directly as flock (cRDF) on large scale water tube boilers for steam or electricity production. The use of cRDF represents a significant cost saving over the production and distribution of dRDF. When integrated with materials recovery for recycling and the production of a low value compost, economic projections indicate that waste segregation can compete on equal terms with other waste disposal options such as mass burn incineration and the use of transfer stations and long haul transportation to landfill sites remote from the area of waste production.

8.29 One area where the densification option for RDF and other wastes is still of interest relates to gasification. Additives such as lime can be intimately mixed and held within the pelletted waste fuel and testwork under gasification conditions has indicated large reductions (over 70%) in acid gas emissions with chlorine and sulphur being retained in the ash.

8.30 Integrated Resource Recovery Facilities already exist overseas. In the UK their development is supported by comprehensive R&D programmes funded by the Departments of Energy and Environment.

Solidification

Introduction

8.31 The term solidification refers to a group of related treatment technologies used to alter the physical and/or chemical characteristics of a waste in order to render it suitable for disposal by landfill. It is generally applied to metal bearing liquors and sludges and hazardous solid wastes such as asbestos.

8.32 The solidification process converts or contains the treated waste constituents with the aim of immobilising them so that they are less able to be re-released into the wider environment. The technique can range from encapsulation of discrete quantities of waste in a jacket of some other material, to fixing by chemical/physical bonds or dispersing the waste in an inert matrix. When the solidified product is to be used for reclaiming land, then sufficient mechanical strength may also be a required objective. Other qualities which the product should exhibit include high resistance to extremes of weather, and to subsequent chemical and biological degradation. To achieve these objectives, reagents or fillers are generally added to the waste to create a material with good structural integrity and with low permeability.

8.33 It may also be necessary to pretreat certain waste streams prior to inclusion in the process. This may include size reduction of solids and chemical treatment of toxic or highly soluble components.

8.34 A number of different techniques for solidification have been developed. These can be summarised by reference to their primary component as:

(a) cement based;

(b) pozzolan based;

(c) lime based;

(d) clay based;

(e) liquid silicate based;

(f) vitrification;

(g) thermoplastic;

(h) thermosetting;

(i) surface encapsulation.

8.35 To date in the UK only cement/pozzolan/lime based solidification processes have been the subject of wide commercial exploitation. Interest in the commercial exploitation of vitrification (a thermal process producing a glass or glass-like product) is, however, increasing, particularly for the treatment of contaminated soils. Other solidification technologies which have not been widely adopted include the following:

(a) Thermoplastic materials which are organic polymers which soften and harden reversibly on heating and cooling. The most commonly suggested thermoplastic materials for waste solidification are asphalt and bitumen. The waste is dried, heated and dispersed through the heated plastic matrix. The mixture is then cooled, solidified and usually disposed of in a secondary container.

(b) Thermosetting plastics which form hardened cross-linked polymers irreversibly on the addition of a catalyst. The urea-formaldehyde solidification process is the most common organic polymer technique, although use of polyester and polybutadiene processes has also been explored.

(c) Surface encapsulation involving the use of an inert material to provide a seal between the waste and the environment. Materials such as polyurethane, epoxide and fibreglass reinforced resins, or a mixture of these, are brushed or sprayed either directly onto the waste particles or onto a container of waste.

(d) Self-cementing systems where a portion of certain waste sludges containing calcium sulphate (e.g. flue gas desulphurisation wastes) or calcium sulphite are calcined then mixed with the remainder of the sludge to produce a partially dehydrated cementitous material suitable for landfilling.

(e) Clay based systems which have been explored as potential fixation agents (with setting agents such as cement) since certain clays have the ability to exchange with, and hence immobilise, certain metal ions and to a lesser extent, polar organics.

Cementitious solidification

8.36 Cement is an anhydrous clinker formed from lime, silica, alumina and iron. The chemistry of cement is complex, but when cement is mixed with water, a hydration reaction occurs, forming a rigid matrix. The strength of the product is related to the degree of hydration, with control of water content consequently being particularly important.

8.37 Cement-based solidification systems are the most developed for the treatment of liquid, sludge and solid wastes. Incoming wastes are generally subject to pretreatment (i.e. oxidation of cyanides, reduction of hexavalent chromium) and dewatering to minimise volume prior to addition of process additives such as flyash and cement. However, the procedure for assessing the suitability of particular waste for treatment has primarily been empirical, involving assessment of a range of waste mixtures and proportions.

8.38 In the UK, there have been three major commercial solidification operations based on cement. One plant, based on a patented cement/sodium silicate process was operated during the 1970s/1980s, but is currently not operational. The two other processes are based on a cement/flyash (pozzolanic/silicate) material. One plant is based in the West Midlands and commenced operations in about 1980 with a plant capacity in the region of 120,000 tonnes per annum (tpa). The second plant, in Essex, came on stream in January 1978, with a capacity of 400,000 tpa.

8.39 A number of problems have been reported with the cement process with regard to the attainment of suitable strength or cohesion of the solidified product. Research is currently aimed at the provision of protocols for assessing the long term integrity of solidified wastes. Most available data suggests that current cement-based systems are best suited for certain inorganic, particularly metal-bearing wastes, where pretreatment can be utilised to ensure low solubility of such species.

8.40 It is known that small amounts of some compounds can significantly reduce the strength and containment characteristics of cement-based solidified wastes. This can be in the form of retardation of setting, such that waste containment is prevented or impaired, or accelerated setting ("flash" set), where inadequate mixing of waste or reagents results. Components of the waste may react leading to swelling or subsequent disintegration of the solidified mass after setting. Interaction between waste components can give rise to accelerated leaching of one or more waste components. Examples of some waste components which may adversely affect cement-based solidification are given in Table 8.3. Conversely, some waste streams may contain compounds known to improve the setting of cement-based systems.

8.41 It is therefore essential that operators closely monitor the materials incorporated into the process to ensure that levels of retardant materials in the product are carefully controlled.

8.42 In general, the incorporation of significant quantities of organic materials into cement-based solidification processes is not recommended since they may interfere with the setting process. Moreover, leaching tests have shown that existing cement-based solidification processes may be ineffective in retaining more than minimal amounts of organic materials. Research is under way to examine the use of various additives to improve the performance of the process with respect to organic materials.

Quality Control

8.43 The initial consideration in the quality control of solidification processes is to ensure suitable test procedures and quality criteria are established for different stages of the operation. The main difficulty lies in defining procedures to predict the potential long-term field performance of a solidified product at the time of preparation, something which is currently not always possible. This is presently the subject of research work. The difficulties are associated with the relatively long period of cement hydration and as a consequence strength development. In the construction industry it is standard practice to determine strengths after 28 days and set a target for strength accordingly.

8.44 Historically, operational practice has not allowed for identification and recovery of those deposits which have subsequently been found to fail a 28-day compressive strength test. In addition, it is generally accepted that no one test is suitable to determine the physical and chemical integrity of the product and an integrated testing programme with a hierarchy of tests is required. Failure at any stage of the tests means that the mix is not acceptable and should be recovered for reprocessing.

Table 8.3 Waste Constituents for which Cement-based Solidification has Limited Application

Waste Constituent	Disadvantage
Highly soluble metal or salt content eg As, Ni, soluble salts of Mn, Sn, Zn, Cu, Pb, Na.	Affect cement setting and may be readily leached
High toxic anion content eg borates	More easily leached than cations
High content of cement setting retarders eg sulphates, halides	Reduce strength of final product
Toxic or nuisance producing substances with odours not destroyed during processing	In plant health hazard/nuisance
Flammable/explosive materials	In plant safety hazard/nuisance
Biologically active compounds eg insecticides, pesticides	May leach readily from final products
Compounds inhibiting cement setting eg sugar, lignite silt	Reduce strength of final product
Production of toxic/hazardous gases on contact with water eg metal carbides	In plant health hazard
Compounds not retained in cement readily eg phenol	Leaches readily

8.45 There are a number of physical, chemical and microstructural tests available which may be appropriate for testing the product at the different stages of operation. They relate to the following product characteristics:

(a) supernatant formation; supernatants initially form above all solidified products but would normally be expected to be rapidly reabsorbed into the setting product;

(b) rate of setting; which can be used as a guide to the eventual strength of a solidified product;

(c) permeability; a critical factor governing the degree to which groundwater or other water percolating into a solidified product may have the potential to leach contaminants;

(d) compressive strength; the products of solidification should resemble solid highly cohesive cementitious materials rather than less cohesive more permeable soil-like materials;

(e) leaching; the leaching characteristic of solidified wastes is very important when assessing the behaviour in a landfill and is subject to much discussion.

8.46 The appropriate minimum criteria for acceptability of chemical species in leachates will depend to a large extent on the arrangements made at the disposal site for leachate collection, treatment and disposal, and will need to take into account whether groundwater or surface water protection is necessary. These considerations will be locally determined in every case, and it is to be expected that the appropriate water company and the National Rivers Authority will advise on these criteria to their own satisfaction during the statutory consultation process.

CHAPTER 9

Evaluating Waste Management Options

Introduction

9.1 This paper has discussed the main options for waste treatment and disposal that are currently available. Selection from this range of choices of the most appropriate method for any area requires decisions to be taken based on knowledge of a large number of variables for a given situation. There is no single perfect option for all circumstances. Therefore this chapter discusses the guiding principles for selection of the options. These can be grouped under the headings of: legislative requirements, environmental impact, economics, flexibility and robustness and the interaction with waste production patterns. There are consequent factors related to planning for waste disposal both in the context of waste disposal plans under the EP Act and land use planning under the Town and Country Planning Act 1990 (para 2.8 et seq). Thus the adoption of a particular strategy should be the result of a careful consideration of these factors which meets the test of providing the Best Practicable Environmental Option.

Legislative requirements

9.2 Optimisation of the methods and techniques to be used is not an easy task in a field where pressure for standards to be raised is growing through the development of a comprehensive array of statutory controls. Public confidence in waste disposal has to be gained by demonstrating environmentally sound practice proven by monitoring data.

9.3 Developments in legislation are determining the agenda for the 1990s requiring that rigorous performance criteria are set and maintained for the future. It will become more important for treatment systems to be secure and capable of being upgraded to meet new criteria.

9.4 The principles for determining the best route for waste management are embodied in European law. The Treaty of Rome which seeks the harmonisation of standards and fair competition within EC member states requires high environmental standards to be set and maintained. Alongside that formal legal framework the EC Strategy on Waste Management sets out a hierarchy of preferred solutions, which requires that waste management starts with the principle of waste minimisation, with landfill as the final option.

9.5 The EP Act has embodied these principles by reinforcing the system of waste regulation and the commitment to raising standards through requirements for operators of sites to be technically competent, to be financially viable and for strict controls on post completion care of sites. The law on the duty of care requires producers of controlled waste and those involved in managing its disposal to ensure that the appropriate standards are met. The White Paper "This Common Inheritance" also promised that further guidance on standards would become available through the series of Waste Management Papers published by the Department of the Environment.

9.6 The new requirement to prepare waste local plans imposed by the Planning and Compensation Act 1991 provides a clearer framework for planning for waste disposal by local planning authorities. In preparing their waste local plans and waste policies, authorities are required by regulation to have regard to any waste disposal plan for their area prepared under EP Act and justify any inconsistencies between their waste policies and the waste disposal plan. Planning Policy Guidance (PPG) note 12 provides advice on this and the forthcoming PPG note on Pollution Controls and Planning will provide further guidance on planning considerations relating to waste disposal and the preparation of waste local plans and policies.

Quality management

9.7 Whatever option is selected, a system for quality management of an operation should be an integral part of the process. This does not necessarily mean that all parts of the strategy are Quality Assured but the investigation of options should have been subject to a quality audit to ensure that fair comparisons had been made and a proper balance achieved between the environmental costs and benefits and the financial costs and benefits.

Flexibility

9.8 Short term solutions should be avoided. It is unlikely that any strategy that does not provide for a secure environmentally sound system of waste management over a ten year period will be satisfactory. This does not mean that only one option or even one site should be chosen but the transition between methods or routes or the development of a strategy should be programmed from the start and the break points planned in advance.

9.9 It is important that the methods and technologies chosen also incorporate some flexibility so that changes in markets or disposal routes can be accepted without disruption to the process of dealing with the continuing generation of wastes for disposal.

9.10 Waste minimisation at the production stage and the Duty of Care on producers and handlers of waste are fundamental parts of this strategy.

Robustness

9.11 Attention to the robustness of the strategy should bring into focus the long term prospects of the techniques with respect to their ability to meet standards of environmental control and in the long term capacity for improvement to account for stricter controls in the future.

9.12 Sites needing Waste Management licences under the EP Act are subject to meeting criteria for certificates of completion and will have to be engineered to a high standard and should be capable of meeting Quality Assurance criteria to satisfy long term future requirements.

Environmental Impact and costs

9.13 An even-handed approach to considering the relative merits of waste management options should cover both the potential environmental impacts and the costs of the choices. For example, guiding criteria for landfill routes would include an assessment of the operational standards of the landfill site as well as the recovery of energy by utilisation of landfill gas, and would be compared with the cost of incineration and the energy consumed and financial cost of longer transport haul. A distant (secure) landfill site may need to be served by bulk haul transfer (road or rail), which also offers local flexibility in the choice of long term landfill as well as the possibility of separation for recycling. These are not new or original concepts but they do need to be applied rigorously to consideration of the proposed waste management option.

9.14 There can be no justification for going for a cheaper option if the long term security of the route is in doubt or aftercare cannot be guaranteed.

Financial and economic appraisal

9.15 The evaluation of the costs of waste treatment and disposal has to be carried out on an equitable basis. Alternative routes should be costed using as far as possible actual estimates and realistic assumptions about market values of recovered products. Where the viability or comparative value of a project hangs on income, say from energy recovery, the security of both the energy sale and the disposal route should be determined such that interrelationships between the various contracts can be maintained to ensure continuation of the scheme.

9.16 Some sensitivity testing of the outcome should be conducted to determine the impact on costs of all the variables.

9.17 The cost of all the elements involved in the process from collection to transfer, treatment and final disposal needs to be considered.

9.18 A similar process was recommended to be carried out when evaluating the financial aspects of recycling schemes (Waste Management Paper No 28 - Recycling). The remainder of this section is based on that account.

9.19 An analysis should take account of the costs associated with existing waste collection and disposal schemes. These would normally include the following:-

(a) the labour costs of collection, delivery and disposal;

(b) the administrative costs of collection, delivery and disposal;

(c) site purchase costs;

(d) the capital costs of equipment;

(e) the operating costs of equipment: fuel, maintenance, insurance, etc;

(f) the disposal costs: landfill, incinerator fees, etc;

(g) the provision of after-care of landfill sites.

9.20 The costs of any recycling schemes that may be associated with the overall strategy will also include:-

(a) the labour and administrative costs of collecting and delivering recyclables either to a market or a processing facility;

(b) any residual disposal costs;

(c) the labour and administrative costs of processing;

(d) the capital costs of the recycling equipment, notably vehicles, materials, handling and processing facilities;

(e) the operating costs of recycling equipment, notably fuel, maintenance, insurance for vehicles and central facilities;

(f) the administrative costs of the recycling scheme, including the costs of promotional, advisory and materials marketing work.

9.21 The revenues and savings from a recycling scheme include:-

(a) the revenue from the sale of the materials;

(b) any revenue from commercial sponsorship of the scheme.

9.22 The costs of the normal collection and disposal service may be referred to as the first (or 'base') option, the impact of any credits available and any other savings in waste handling costs should already be contained within the calculation.

9.23 The process of financial appraisal should take into account both immediate and future costs and benefits, using discounting as the means of comparing costs and benefits which accrue at different times. Discounting requires a discount rate. While central government does not prescribe what discount rate local authorities should adopt, the Government itself might in similar circumstances use a real rate (i.e. a rate excluding the rate of inflation). A time period of at least 10 years should be used and up to 20 years is generally defensible.

9.24 Many benefits only accrue once a system is up-and-running (the exception may be large metropolitan authorities where local availability of landfill sites means that waste disposal costs are already high). For example, current evidence suggests that cost savings for recycling schemes from household waste will occur only after a scheme has become established and households are recycling 20-30% of their waste.

9.25 All of these figures will be subject to uncertainty. The costs of collection, separation, processing and disposal are dependent on energy costs which fluctuate with supply and demand; selling prices for recycled materials and market prospects change considerably. An appraisal should take these uncertainties into account by way of a sensitivity analysis that identifies the effect such changes would have on the overall viability of the project, having considered a plausible range of values for each uncertain variable and the effect of the analysis on the ranking of various options. This analysis should be carried out for all major variables. As a result of the sensitivity analysis, authorities should get a better idea of the viability of a particular strategy.

9.26 A guide which shows practitioners how to conduct an appraisal and a sensitivity analysis is *"Economic Appraisal in Central Government"* by H M Treasury (HMSO 1991). Further information on environmental impacts and cost-benefit analysis is also contained in the recent DoE publication "Policy Appraisal and the Environment".

9.27 A financial analysis is concerned only with costs and revenues which accrue to the authority itself. An authority which is evaluating options should be aware that its policies are likely to give rise to other costs and benefits which fall on the community at large. These should also be assessed.

9.28 When considering the results of the financial appraisal, authorities should keep in mind these wider costs and benefits. They are unlikely to be able to attempt a full cost-benefit analysis (by putting values on them) but it will be helpful to list the wider environmental impacts and to consider whether they tip the balance of the decision so that the authority may be willing to bear a financial loss in order to generate wider benefits.

APPENDIX 1

The Department's Controlled Waste Research Programme

1. Controlled waste management research is carried out to guide and inform policy decisions related to waste management in the U.K., to support those policies within the European Community and internationally, to monitor developments in waste disposal practices, to assess both the long- and short-term environmental implications of such practices and to provide the data for guidance on best practice in the waste management industry.

2. This research programme works towards the following policy objectives:

(a) To ensure effective and economic controls over the disposal of industrial, commercial and domestic waste and to improve general standards of solid waste management.

(b) Subject to safe environmental treatment to minimise the international trade in waste.

(c) To ensure that industrial, commercial and domestic waste is minimised at source, recycled where practicable and otherwise disposed of safely within the U.K.

3. Much of the Department's research in this area is published as Waste Management Papers which are updated and added to as necessary. This research also contributes to national and international advisory and working groups. This includes work on composting, sampling and analysis, dioxin limits and landfill engineering. Where the work is appropriate, the results are presented as papers at major conferences and seminars, or published as Departmental reports. As well as commissioning original work, the Department also periodically undertakes reviews of the available international knowledge in major areas of interest to minimise duplication, and to ensure its research programme meets the above objectives effectively.

Scope of research

4. The research programme is divided into four main areas:

(a) Materials Processing - this includes recycling, composting and combustion/incineration.

(b) Industrial Waste Management - this covers problem wastes, excluding the Department's Radioactive waste research. Currently this includes work on PCBs, asbestos, List 1/Red List substances.

(c) Landfilling Practices - the continuing investigation of landfill processes and associated problems such as leachate, landfill gas, health and safety, site stability, etc.

(d) Long-term Monitoring - this area concentrates on providing information on the long-term effects and stability of landfilling and solidified wastes.

Current objectives

4.1 Materials Processing- Current Objectives

(a) To foster efficiency in waste management.

(b) To investigate the environmental acceptability of the various materials processing options which form alternatives to landfilling.

This is to be achieved by:

For composting - reviewing the status of composting technologies worldwide, determining specifications for compost material, determining the size and value of potential markets and evaluating emissions from the conversion of waste to compost.

For incineration - much work has been done in evaluating safe operating practice, equipment and sampling. Work continues on these, particularly on the treatment of residues requiring final disposal.

Recycling - assessing market potential for recycled materials and the potential of centralised resource recovery schemes.

Efficiency - studying the composition of waste, the economic and environmental costs of the various disposal paths and the impact of policies, legislation and regulations.

Generally - reviewing waste sorting and processing technologies and waste disposal overseas. Assessing the health and safety aspects of waste disposal practices.

4.2 Industrial Waste Management - Current Objectives

(a) To assess disposal options for selected problem wastes.

Plate 22 Test rig for research into the properties of compacted landfill waste *(DOE)*

(b) To investigate and develop technologies for the physical, chemical and/or microbiological degradation, detoxification or immobilisation of such wastes and to review the environmental impact of the processes;

(c) To provide data on background levels and distribution of selected problem wastes.

This is to be achieved by :

Degradation/toxicity testing - continuing the development of an anaerobic treatability test and investigating the treatability of selected problem wastes such as PCBs, PCNs, etc.

Disposal options - continuing work on alternative methods of disposal for asbestos, PCBs and other List 1 substances, their fate and the long-term stability of solidified waste forms. Also, the disposal of incinerator residues from MSW, sewage sludge and power generation.

Generally - investigating the historical background levels of PCBs, the environmental impact of PCB substitutes and the potential flux of PCBs, vinyl chloride monomer etc. and their degradation products. Also, to develop a categorisation scheme for non-hazardous industrial wastes, assess the safe disposal of clinical wastes and investigate waste streams in an attempt to identify emerging waste problems.

These results and any necessary additional work, will support EC legislation such as the Framework Directive on Waste and the Hazardous Waste Directive. Waste Management Papers benefiting from revision will include 6, 23 and 25 and other technical guidance which may required.

4.3 Landfill Practices - Current Objectives

(a) To provide technical guidance which will enable an improvement in landfilling practices as well as optimising the resource.

(b) To ensure that landfilling is conducted as a safe and effective long-term disposal option.

(c) To provide technical guidance on the assessment of landfill pollution potential.

This is to be achieved by:

For landfill processes - statistical characterisation of sites, metal speciation studies, gas potential testing, studies of accelerated stabilisation and other work for certificates of completion, as well as further co-disposal research.

For landfill engineering and reinstatement - studies on engineering, leachate management, completion, landfill capping, recirculation and methane oxidation.

Leachate treatment - studies of the attenuation of pollutants in natural systems, as well as continuing studies of nitrification/ denitrification and the treatment of landfill leachate to surface water standards.

Modelling and risk assessment - the development of gas production and gas migration models and the development of a risk assessment methodology for appraisal of landfill engineering plans.

Surveys and monitoring - studies on the environmental impact and health effects of landfilling, statistical sampling studies and monitoring protocols.

As can be seen above, research subjects in this area are diverse and the list is not exhaustive. However, the output in this area is geared to revision of Waste Management Papers and to the provision of information and guidance for the implementation of the Environmental Protection Act 1990. Technical guidance will be published as necessary and as data becomes available.

4.4 Long-term Monitoring - Current Objectives

(a) The provision of long-term data on landfill leachate migration, attenuation and treatment.

(b) The assessment of the long-term integrity of landfill caps and solidified waste.

Plate 23 Experimental reed bed for polishing treated leachate to surface water standards *(DOE)*

This will be achieved by:

> Monitoring the effectiveness of leachate treatment systems and artificially placed leachate attenuation systems.
>
> Observing the migration and attenuation of components of leachate within groundwater.
>
> Monitoring the percolation through well-characterised landfill capping systems.
>
> Monitoring the ageing process of solidified waste in the field.

The results will be used to provide guidance for implementation of the Environmental Protection Act and updating Waste Management Papers and other technical guidance as necessary.

BIBLIOGRAPHY

Useful references for the reader to gain further information on the waste management industry

Waste Management Papers

WMP 2 Waste Disposal Surveys. HMSO 1976 (ISBN 0117510033).

WMP 3 Guidelines for the Preparation of a Waste Disposal Plan. HMSO 1976 (ISBN 0117511242).

WMP 4 The Licensing of Waste Facilities. Second Edition HMSO 1988. (ISBN 0117521574).

WMP 5 The Relationship between Waste Disposal Authorities and Private Industry. HMSO 1976 (ISBN 0117509205).

WMP 6 Polychlorinated Biphenyl (PCB) Wastes - a Technical Memorandum on Reclamation, Treatment and Disposal. HMSO 1976 (ISBN 0117510009).

WMP 7 Mineral Oil Wastes - a Technical Memorandum on Arisings, Treatment and Disposal. HMSO 1976 (ISBN 0117510602).

WMP 8 Heat Treatment Cyanide Wastes - a Technical Memorandum on Arisings, Treatment and Disposal. Second Edition. HMSO 1985 (ISBN 0117518131).

WMP 9 Halogenated Hydrocarbon Solvent Wastes from Cleaning Processes - a Technical Memorandum on Reclamation and Disposal. HMSO 1976 (ISBN 011751103X).

WMP 10 Local Authority Waste Disposal Statistics 1974/75. HMSO 1976 (ISBN 011751120X).

WMP 11 Metal Finishing Wastes - a Technical Memorandum on Arisings, Treatment and Disposal. HMSO 1976 (ISBN 0117511226).

WMP 12 Mercury Bearing Wastes - a Technical Memorandum on Storage, Handling, Treatment Disposal and Recovery. HMSO 1977 (ISBN 0117511269).

WMP 13 Tarry and Distillation Wastes and other Chemical Based Wastes - a Technical Memorandum on Arisings, Treatment and Disposal. HMSO 1977 (ISBN 0117511277).

WMP 14 Solvent Wastes (excluding Halogenated Hydrocarbons) - a Technical Memorandum on Reclamation and Disposal. HMSO 1977 (ISBN 0117511285).

WMP 15 Halogenated Organic Wastes - a Technical Memorandum on Arisings, Treatment and Disposal. HMSO 1978 (ISBN 0117513709).

WMP 16 Wood Preserving Wastes - a Technical Memorandum on Arisings, Treatment and Disposal. HMSO 1980 (ISBN 0117514764).

WMP 17 Wastes from Tanning, Leather Dressing and Fellmongering - a Technical Memorandum on Recovery,

Treatment and Disposal. HMSO 1978 (ISBN 0117513202).

WMP 18 Asbestos Waste - a Technical Memorandum on Arisings and Disposal. HMSO 1979 (ISBN 0117513849).

WMP 19 Wastes from the Manufacture of Pharmaceuticals, Toiletries and Cosmetics -a Technical Memorandum on Arisings, Treatment and Disposal. HMSO 1978 (ISBN 0117513180).

WMP 20 Arsenic Bearing Wastes - a Technical Memorandum on Recovery, Treatment and Disposal. HMSO 1980 (ISBN 0117514721).

WMP 21 Pesticide Wastes - a Technical Memorandum on Arisings and Disposal. HMSO 1980 (ISBN 0117514845).

WMP 22 Local Authority Waste Disposal Statistics 1974/75 to 1977/78. HMSO 1978 (ISBN 0117514535).

WMP 23 Special Wastes - a Technical Memorandum Providing Guidance on their Definition. HMSO 1981 (ISBN 0117515558).

WMP 24 Cadmium Bearing Wastes - a Technical Memorandum on Arisings, Treatment and Disposal. HMSO 1984 (ISBN 01175176X).

WMP 25 Clinical Wastes - A Technical Memorandum on Arisings, Treatment and Disposal. HMSO 1983 (ISBN 0117517194).

WMP 26 Landfilling Wastes - a Technical Memorandum on Landfill Sites. HMSO 1986 (ISBN 0117518913).

WMP 27 Landfill Gas - a Technical Memorandum on the Monitoring and Control of Landfill Gas. HMSO 1991 (ISBN 0117524883).

WMP 28 Recycling - A Memorandum for local Authorities on Recycling - HMSO 1991 (ISBN 011752445).

HMSO offer an 'out of print' service for Waste Management Papers. Tel: 071-873 8455.

This Common Inheritance - Britain's Environmental Strategy: Cmnd 1200: HMSO 1990 (ISBN 0101120028).

This Common Inheritance - The First Year Report: Cmnd 1655; HMSO 1991 (ISBN 0101165528).

This Common Inheritance - summary of White Paper. HMSO - 1990 (ISBN 01175223348).

Pollution of Water by Tipped Refuse: HMSO 1961 (ISBN 0117502790).

Report of the Technical Committee on the Disposal of Solid Toxic Wastes: HMSO 1970 (ISBN 0117502782).

Report of the Working Party on Refuse Disposal, Chairman J Sumner: HMSO 1971 (ISBN 0117503487).

War on Waste: A Policy for Reclamation: Cmnd 5727, September 1974. HMSO (ISBN 0101572700).

Report on the Disposal of Awkward Household Wastes: 1974. HMSO (ISBN 0117507296).

First Report of the Standing Committee on Research into Refuse Collection, Storage and Disposal: Sept 1969-Dec 1972. HMSO (ISBN 0117506214).

Waste Management Advisory Council - 1st Report 1976. Joint DoE/DoI ISBN 0117510076

Occurrence and Utilisation of Mineral and Construction Wastes: HMSO 1991 (ISBN 0117524840).

Report on Waste Paper Collection by Local Authorities - Waste Management Advisory Council Paper No.2.

An Economic Case Study of Waste Oil - Waste Management Advisory Council Paper No 3.

Co-Operative Programme of Research on the Behaviour of Hazardous Wastes in Landfill Sites: HMSO 1976 (ISBN 0117512).

A Health and Safety Commission Guidance Document, "The Safe Disposal of Clinical Waste". HMSO 1982.

Energy from Waste: HMSO (ISBN 011 5122028).

Mechanised Household Waste Sorting in the UK - A General Guide. HMSO (ISBN 011 7516295).

Royal Commission on Environmental Pollution: Eleventh Report: Managing Waste: The Duty of Care. HMSO 1985.

The Future for Rubbish - waste management options for the environment reviewed; by Dr G Bailey and Richard Hawkins; published by the Conservative Political Centre (1983). HMSO.

Department of the Environment. Managing Waste: The Duty of Care. The Government's Response to the Eleventh Report of the Royal Commission on Environmental Pollution (Pollution Paper No.24). HMSO 1986.

Department of the Environment. Planning Policy Guidance: Development Plans and Regional Planning Guidance (PPG 12) February 1992. HMSO ISBN 0117525863

An Economic Case Study of Waste Oil - Waste Management Advisory Council Paper No 3. DOE.

The Fourth Report of the Trade & Industry Committee (Session 1983-84): The Wealth of Waste: HMSO (ISBN 01026484X).

Mechanised Household Waste Sorting in the United Kingdom - A General Guide, HMSO (ISBN 0117516295).

Report of a Review of the Control of Pollution (Special Waste) Regulations 1980. DOE 1985.

House of Lords Select Committee on Science and Technology. Hazardous Waste Disposal: Government Response. HMSO 1985.

House of Lords Select Committee on Science and Technology. Hazardous Waste Disposal: Review of the Control of Pollution (Special Waste) Regulations 1980. HL 234 S 84-5 (ISBN 0104234857).

Her Majesty's Inspectorate of Pollution: Fourth Annual Report 1990-91: HMSO 1991 (ISBN 0117525197).

KNOX, K. (1989) A review of technical aspects of co-disposal (PECD 7/10/214). Department of the Environment Report No CWM 007/89. 260pp.

NATIONAL RIVERS AUTHORITY (1991) Policy and practice for the protection of groundwater, 48pp. (Draft; Final version should be available July 1992)

SEYMOUR, K.J. and PEACOCK, A.J. (1989) Earthworks on landfill sites. Proc. 2nd Int. Landfill Symposium, Sardinia. B.IX-1-10. Centro di Ingegneria per la Protezione dell'Ambiente srl.

Environmental Assessment. Joint Circular DOE 15/88, Welsh Office 23/88. HMSO (ISBN 0117521175)

Policy Appraisal and the Environment. Department of the Environment 1991 HMSO (ISBN 0117524875).

Economic Appraisal in Central Government: A Technical Guide to Government Departments. H M Treasury 1991 HMSO (ISBN 0115600345).

Waste Management-The Duty of Care, A Code of Practice. Department of the Environment 1991, HMSO (ISBN 011752557X)

Hazardous Waste Inspectorate Scotland Report. Scottish Office Library (ISBN 07480 0368 1)

Review of MSW Incineration in the UK, LR776(PA), Warren Spring Laboratory. (ISBN 085624 634 4).

A Review of Mass Burn Incineration as an Energy Source, ETSU-R-57,1990, Energy Technology Support Unit, AEA Harwell.

Code of Good Agricultural Practice for the Protection of Water. MAFF/Welsh Office Agriculture Dept. MAFF Publications.

The United Kingdom Waste Management Industry - J.R. Holmes, Institute of Waste Management, Jan 1992.

UK Legislation

Environmental Protection Act 1990

Control of Pollution Act 1974

Control of Pollution (Supply and Use of Injurious Substances) Regulations 1986, SI No 902. [Replaced the Regulations of 1980, SI. No 638].

Control of Pollution (Landed Ships' Waste) Regulations 1987, SI No 402.

The Collection and Disposal of Waste Regulations 1988, SI No 819.

The Transfrontier Shipment of Hazardous Waste Regulations 1988, SI No.1562.

The Control of Pollution (Special Waste)(Amendment) Regulations 1988, SI No 1790.

Control of Pollution (Landed Ships' Waste)(Amendment) Regulations 1989, SI No 65.

Control of Pollution (Special Waste) Regulations 1980, SI No 1709.

Trade Effluents (Prescribed Processes and Substances) Regulations 1989 SI No 1156.

Control of Pollution (Amendment) Act 1989.

Town and Country Planning Act 1990

Local Government Act 1985

Local Government Act 1988

Local Government and Housing Act 1989

Town and Country Planning General Development Order 1988, SI No 1813 as amended.

Town and Country Planning (Assessment of Environmental Effects) Regulations 1988, SI No 1199.

Town and Country Planning (Development Plan) Regulations 1991 SI No 2794

Environmental Protection (Prescribed Processes and Substances) Regulations 1991, SI No 472.

Disposal of Controlled Waste (Exceptions) Regulations 1991, SI No 508.

Waste Regulations and Disposal (Authorities) Order 1985. SI No 1884

The Environmental Protection (Duty of Care) Regulations 1991 SI No 2839.

Planning and Compensation Act 1991.

Controlled Waste (Registration of Carriers and Seizure of Vehicles) Regulations 1991. SI No. 1624.

The Electricity (Non-Fossil Fuel Sources) (England and Wales) Order 1990 SI No 1859

The Electricity (Non-Fossil Fuel Sources) (England and Wales) Order 1991 SI No 2490

Water Act 1989

Water Resources Act 1991

Electricity Act 1989

Scottish Legislation

The Control of Pollution (Licensing of Waste Disposal) Scotland Regulations 1977. SI 1977 No. 2006.

The Environmental Assessment (Scotland) Regulations 1988 SI 1988 No. 1221.

Refuse Disposal (Amenity) Act 1978.

Town and Country Planning (Scotland) Act 1972 (as amended).

The Town and Country Planning (General Development Procedure) (Scotland) Order 1992.

The Town and Country Planning (General Permitted Development) (Scotland) Order.

Civic Government (Scotland) Act 1982.

Local Government (Scotland) Act 1973.

The Prevention of Environmental Pollution from Agricultural Activities. Scottish Office Agriculture and Fisheries Department.

European Community Legislation

Directive on waste 75/442/EEC amended by 91/156/EEC, OJ No L78, 26.3.91, p32.

Directive on toxic and dangerous waste 78/319/EEC, OJ No L84, 31.3.78, p43.

Directive on the protection of groundwater against pollution caused by certain dangerous substances, 80/68/EEC, OJ No L20, 26.1.80, p43.

Directive on the assessment of the effects of certain public and private projects on the environment 85/337/EEC, OJ No L175, 5.7.85, p40.

Directive on the supervision and control within the European Community of the transfrontier shipment of hazardous waste 84/631/EEC as amended by 86/279/EEC, OJ No L181, 4.7.86, p13.

Directive on the prevention of air pollution from new municipal incineration plants 89/369/EEC, OJ No L163, 14.6.89, p32.

Directive on the reduction of air pollution from existing municipal waste incineration plants 89/429/EEC, OJ No L203, 15.7.89, p50.

Council Resolution on waste policy 90/C 122/02, OJ No C122, 18.5.91, p2.

Directive on hazardous waste 91/689/EEC, OJ No L377, 3.12.91, p20.

Proposal for a Council Directive on the landfill of waste 91/C 190/01, OJ No C190, 22.7.91, p1.

Printed in the United Kingdom for HMSO Dd 295813 C50 4/92 59226